Biological Treatment Process In Drinking Water

Biological Treatment Process In Drinking Water

Shraddha Gautam

ARCLER PRESS

www.arclerpress.com

Biological Treatment Process In Drinking Water

Shraddha Gautam

Arcler Press
224 Shoreacres Road
Burlington, ON L7L 2H2
Canada
www.arclerpress.com
Email: orders@arclereducation.com

ISBN: 978-1-77407-791-7 (Hardcover)

ABOUT THE AUTHOR

Shraddha Gautam is an accomplished Sr. Scientist, at Advancells, with 16 years of onsite experience, mainly in the therapeutic application of regenerative medicine and regulatory support. She has a proven ability to foster technological advancements and quality excellence. She has been a part of India's top most institutions facilitating stem cell research and therapeutic applications and has gained extensive experience in the field of stem cell biology, its technological advancements and clinical implementation. She has acquired in-depth understanding of regulatory parameters required to be fulfilled for smooth implementation of science in clinical practice. Embracing the core values of integrity, growth and compassion; Shraddha is continuing her contributions to the field of stem cell technology and cellular therapies by presenting many research publications in renowned medical journals.

TABLE OF CONTENTS

LIST OF FIGURES

PREFACE

Since the beginning of 20th century, humans have utilized microbial biomass to degrade nutrients, chemicals, and other organic substancesin water that can be harmful to living beings. Although, use of the biological methods, for water treatment has been very limited in the past; recently, due to technological advancements biological methods of treating water have been acknowledged as more viable and effective option. These advancements in biological treatments is essential due to various reasons, some of them can be noted as:

(I) Growing expenditures and increasing complications of managing water-treatment residuals (i.e., membrane concentrates)

(II) The advent of novel contaminants that are predominantly responsive to biological degradation (i.e., perchlorates)

(III) The drive for green treatment technologies (i.e., methods that proficiently degrade pollutants instead of accumulating them)

(IV) Regulations and laws that are further responsible for preventing the development of disinfecting by-products (i.e., DBPs)

(V) The development of membrane-based water treatment processes, which are highly vulnerable to biological fouling.

This book provides a brief overview of the basic processes of biological water treatment, along with the description of mechanisms for modeling and integrating biological stages into different industrial processes. Further to which, the book essentially covers the physical, biological and chemical basics, including microbial metabolism, wastewater characteristics, wastewater treatment processes (including aerobic and anaerobic treatment processes). It also includes detailed information on membrane technology, technologies responsible for swift removal of micropollutants, activated sludge processes, membrane bioreactors (MBRs) and the evaluation and selection of different aeration systems.

Overall, the book consists of eight chapters. Chapter 1 introduces the readers with fundamental concepts of biological water treatment. The fundamental concepts of microbes and ecological aspects of water are essential in understanding the biological wastewater treatment. Chapter 2 focuses on microbiology and ecology of wastewater treatment. Chapter 3 details the basics and applications of aerobic wastewater treatment. Chapter 4 is all about activated sludge treatment processes and industrial applications.

Chapter 5 is more about the anaerobic water treatment processes and their applications. Membrane bioreactor technology is a leading water treatment technique, which deals with hybrid treatment approaches (i.e., membrane processes and biological treatments). Chapter 6 deals with membrane bioreactor technology and its application in treatment processes. Chapter 7 deals with aerobic membrane bioreactor technologies for wastewater treatment. Finally, Chapter 8 discusses the anaerobic membrane bioreactor technologies for treating drinking water.

This book is an essential read for students and teachers in the field of water treatment technologies. Moreover, this book can be considered a ready reference for engineers, microbiologists, chemists, environmental officers, consultants and agencies, in both industry and academia.

Author

CHAPTER

1

BIOLOGICAL WATER TREATMENT

CONTENTS

1.1 INTRODUCTION

Ecological instability can be experienced as a result of the rising population, unsystematic use of resources, man's activities and growing pollution on national and international levels. With the rising issue of disposal of large quantities of unprocessed matter, in the water; it has been quite evident that the amount of oxygen in the receiving water is getting reduced, due to accelerated microbial activity. Increased patterns of consumption have caused the production of a high amount of wastewater with high organic composition. As a matter of fact, the presence of trace contaminants and reduced oxygen levels in these effluents can have adverse impacts on the environment, accompanied by decreased environmental prosperity and biological variety. These detrimental effects have fuelled the desire to understand the impact of pollution on the bodies of water, and also establish the appropriate processed required to scale down these emissions and treatment practices (Bouwer& Crowe, 1988; Scott &Ollis, 1995).

Today, various manmade technologies are helpful in treating waste. The choice of ideal technology, further depends upon the water source and goals associated with the same; which can further be understood by noting following parameters:

1) Industrial wastewater, produced by industries, while carrying out their regular activities; and is mostly depend on the local sanitation structure for processing waste
2) Residential wastewater generated from toilets, gray water, and other household activities.

Providing high economic advantages, biological wastewater treatment processes are very effective, particularly when associated with resource recovery and waste stabilization (Servais et al., 1994; Andersson et al., 2001). Thus, based on communal behavior, water usage, the type of wastewater accumulation structure, climatic situation, the amount and kinds of businesses in a catchment and the nature of the sources; the contents of wastewater, including its components and concentration, differs significantly from place to place. Keeping in mind this diversity, wastewater treatment methods should naturally be flexible; however, on occasions they must also be designed according to the certain situation and waste. This chapter aims to explain the various practices of biological treatment, and address their respective capabilities to treat different wastes; depending on the qualities of the waste and need for the resource revival (Otte& Rosenthal, 1979; Benner et al., 2013).

1.2 ALTERNATIVES FOR WASTEWATER TREATMENT

Considering its importance, waste has been treated and has been paid special attention for the past thousands of years. However, a modification in understanding has brought changes in its management. Drafted by Aristotle in the 4th century B.C, the *Athenian Constitution* specified ways for proper management of waste (Seghezzo et al., 1998). Since, the connection between man's health and waste had not been established back then; attention was probably focused on the effluvium and physical appearance. Apparently, the development of technologies and strategies for handling waste to deal with health issues, started when the relationship between health and waste became more noticeable (Chernicharo, 2006; Deegan et al., 2011).

Advancements in treatment technology i.e. biological, physical and chemical processes, have been gradual; several of these are still being utilized in various fields. Chemical procedures treat impurities by provoking certain chemical reactions or by the addition of chemicals. Some general examples of chemical treatment methods are disinfection, precipitation, and adsorption. While gas transfers, sedimentation, mixing, flotation, flocculation, filtration, and screening are also some of the examples of physical methods that focus on the utilization of physical forces (Machado et al., 2007). In an industrial setting, combinations of physical and chemical processes are mostly used together for treatment. As compared to physiochemical processes, biological methods disinfect organic adulterants (e.g. biodegradable organic substances) commonly and also more effectively through microbiological activity; some of these approaches that are more prevalent can be noted as trickling filters, activated sludge plants, ponds treated with aerobic reactor systems (Tchobanoglous, 2003; Lettinga, 1995; 1996) (Jewell, 1996).

The selection of treatment can also be directly dependent upon analyzing some prominent characteristics of wastewater like concentration, source, and rate of its production. The system that is being used for the wastewater treatment should be proficient enough to reduce energy consumption; along with being an ideal system to omit adverse treatment outcomes and eliminate basic adulterants. Thus, planning and execution of an ideal wastewater treatment process is a crucial task; and may require the support of further elaborative techniques as well. ,In any manufacturing setup, implementation of treatment requires specific demands; depending upon the distinct objectives of the treatment. For instance, domestic wastewater aims to minimize the organic adulterants in a system, and further increase number of secondary nutrients (phosphorus, P; nitrogen, N). On the other hand, physical treatment may be a necessity, before the application of biological processes to cleanse solids; and also administration of soluble organic substances in a waste. Industrial wastewaters eradicate and decrease the number of organic compounds; mainly the contaminants that exist in some wastes. Thus, \ chemical procedures should also be included in industrial treatment methodology. However, biological approaches are used whenever possible (Gautam et al., 2007; Ozgun et al., 2013).

By the unified action of microorganisms, such as protozoa, rotifers, algae, bacteria, and fungi; the biological deterioration of organics can be attained. In this regard, it is very important to note that standards for the maximum level of unacceptable substances that exist in discharged water have been set down by administrative ministries to preserve ecological stability in the receiving water. The processes mentioned below are generally executed in regular wastewater treatment plants, to attain the appropriate level of emission; before they can be carefully discarded into the receiving water (Balkema et al., 2001; Buyukkamaci&Koken, 2010).

1.2.1 Preliminary Treatment

Pretreatment is required to secure the pumping equipment and assist in achieving the succeeding treatment steps. This can be explained as, to remove extra grease and oil in the discharge, froth flotation is used occasionally;while the elimination of heavier compounds or bigger floating or suspended solids that can harm pumps is accomplished by pretreatment devices like screen and/or grit removal systems.

1.2.2 Primary Treatment

Most of the solids that can be easily settles are removed from the wastewater by simple sedimentation; which is a purely physical process. The technique is primarily used to extract solids, which can easily settle down in wastewater. In this process, the horizontal velocity of the water through the settle is maintained at a level that provides solids adequate time to settle and floatable material can be removed from the surface. Therefore, primary treatment steps requires settling tanks, clarifiers, and/or flotation tanks; which can further send separated solids to digestion units, and supernatants to subsequent, typically microbiological, treatment units.

1.2.3 Secondary Treatment

The secondary treatment is further required for biochemical disintegration of organic material in wastewater; through effective utilization of microbial communities, with different growth environments. A wide range of reactors are used for this treatment, such as fixed-film reactors, pond or lagoon systems, biofilm, and suspended biomass.

1.2.4 Secondary Clarification

By transforming waste carbon into new cells, many biological treatment procedures generate surplus biomass. Solids must be removed from the secondary treatment discharges, before the last treatment steps, like nutrient removal or disinfection is achieved; either through settling or by using the membranes..

Thus, based on the type of digestive system used, the extracted solids are either transported to digesters for purification of solids and removal or recovered to the head process chain.

1.2.5 Tertiary Treatment

To attain a greater emission level than obtained by the regular secondary treatment processes, the tertiary or advanced treatment is used. It includes ion exchange, nutrient removal, activated carbon adsorption, electrodialysis, chemical oxidation, and reverse osmosis; as some of the finishing processes.

Exclusively designed for nullifying residual organisms present in wastewater after all the past treatment procedures, the ultimate effluent disinfection is mostly carried out after the secondary or tertiary treatment by ozonation, chlorination, ultraviolet methods, and similar other techniques; however, technically speaking this step is not a part of tertiary process.

1.3 BIOLOGICAL TREATMENT POSSIBILITIES

In a treatment system, various predominant microorganisms demonstrate basic metabolic pathways; biological processes are categorized based on these pathways. Further to which, depending upon the oxygen available and used, these processes are categorized as anaerobic, aerobic and anoxic (Ho et al., 2012; Schoels et al., 2012).

1.3.1 Aerobic Procedures

Aerobic processes are treatments that take place in the presence of molecular oxygen (O2) and utilize aerobic respiration to produce cellular energy. Even though, they result in greater residual solids as cell mass, they are highly metabolically active.

1.3.2 Anoxic Processes

They produce energy through anaerobic respiration and take place in the absence of free molecular oxygen (O2). As their terminal electron acceptor, microorganisms utilize combined oxygen from inorganic substances in the waste (e.g., nitrate). By denitrification anoxic procedures are average biological nitrogen removal systems.

1.3.3 Anaerobic Processes

Resulting in methanogenesis and sulfate reduction these processes take place in the absence of free or combined oxygen. Through treatment, they commonly produce fewer amounts of biosolids and biogas (i.e., methane) as a beneficial by-product (El-Gohary et al., 1995).

Biological wastewater treatment processes are also categorized based on the growth conditions in the reactor, other than their classification according to microbial metabolism and/or oxygen utilization (Fig. 1.1). Attached growth processes and suspended growth are the two main classes, in this case.

1.3.4 Suspended Growth Processes

Microorganisms responsible for the transformation of organic waste substances to biomass and simpler compounds are conserved in a suspension, within the liquid phase; in these processes. There are various kinds of anaerobic and aerobic suspended growth processes. Sequencing batch reactors, activated sludge and aerated lagoons are the procedures, included in aerobic processes; meanwhile, plug-flow digesters, stirred-tank reactors, bag digesters and baffled reactors with organisms are mainly suspended in the liquid phase (Ludolph et al., 2009; Hargreaves et al., 2013).

1.3.5 Attached Growth Process

The microorganisms in these processes are self-immobilized on granules or flocs or can also be present on the surfaces (e.g. inert packing materials, stones) the system to decompose waste. Attached growth processes may be either anaerobic or aerobic. Upflow anaerobic sludge blanket (UASB), various hybrid anaerobic reactors (HAR), anaerobic rotating biological contactors, down-flow packed bed reactors, anaerobic fluidized bed reactors are included in anaerobic systems. At the same time, packed-bed reactors, trickling filters, rotating biological contactors, and roughing filters are included in aerobic processes. For the anaerobic treatment of domestic and industrial wastewater UASB reactors are more commonly used (Shipin et al., 2007; Arslan et al., 2014).

1.4 TREATMENT PROCESSES FOR AEROBIC BIOLOGICAL WASTE

The formation of simple organics, new cell biomass, and carbon dioxide is achieved, after the basic aerobic waste treatment systems provide a position; where microorganisms are subjected to molecular oxygen (O2) to oxidize compound organic substances, existing in the waste. In developed countries, the most frequently used and renounced biological treatment process is the activated sludge process (ASP) (Lawrence & McCarty, 1970; Ponsá et al., 2008).

1.4.1 Activated Sludge Process

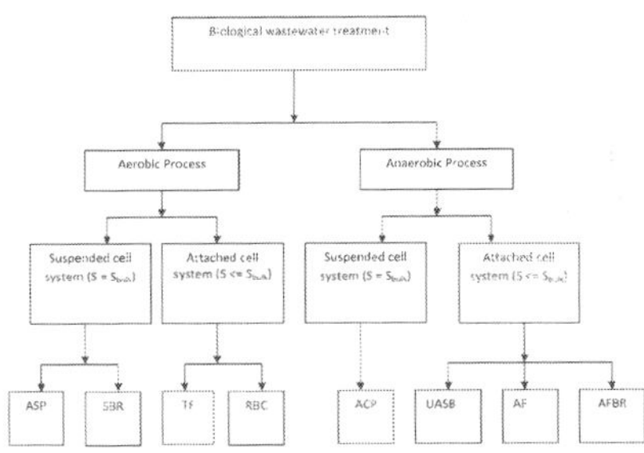

Figure 1.1: Various biological treatment processes.

Note:- *S bulk, substrate concentration in the bulk of the liquid; UASB, up-flow anaerobic sludge blanket; SBR, sequencing batch reactor; S, the concentration of substrate accessible to microorganisms; RBC, rotating biological contactor; ACP, anaerobic contact process; AF, anaerobic filter; ASP, activated sludge process; AFBR, anaerobic fluidized bed reactor; TF, trickling filter*

[Source: https://www.researchgate.net/publication/234167418_Wastewater_treatment_Biological]

The discovery of the activated sludge processes took place at the start of the 1900s. The word "activated" signifies solids that are responsible for the decomposition of waste; it was later realized that the "activation" part of the sludge was an intricate combination of microbes. Further to which, including resident organisms and wastewater; the liquid inactivated sludge systems is known as the "mixed liquor". It should be noted that aerobic suspended cell systems are standard ASPs. Depending on the framework of the process, mineralization of waste organic compounds is complemented by the elimination of inorganic compounds like phosphorus and ammonia; or the production of new microbial biomass (Zhou et al., 2006; Chowdhury et al., 2010).

There have been numerous manifestations of the ASP. Typically plants include continuous-flow stirred-tank reactors and standard step aeration; consisting of a secondary clarifier, an aeration tank, and common pretreatment processes, a conventional ASP has been depicted in Fig. 1.2. For the microorganisms to grow, the aeration tank delivers a steady supply of dissolved oxygen through subsurface or surface aerator. The residing microorganisms utilize the organic matter present in the wastewater as it moves through the tank. The microorganisms are separated from the effluent flowing through the aeration tank with the help of a clarifier. The clarifier supernatant is transported to the treatment or disinfection section and finally released into the receiving water. Biosolids from the settler are either moved to the digesters for additional processing or recovered back to the start of the treatment system.

1.4.2 Aeration Tanks

Exposed to the atmosphere, aeration tanks are mostly built uncovered. Two basic methods are used to supply air to the microbes in the aeration tank: either using diffusers or mechanical aerators. Mechanical aerators like brush aerators and surface aerators encourage the diffusion of oxygen to water from the atmosphere and oxygenate the water's surface. By regulating the rotor's speed, the quantity of dissolved oxygen in the liquid can be standardized. However, due to the greater amount of oxygen transfer efficiency, diffusers are mostly chosen, as they bubble air right into the tank at depth. In aerobic biological wastewater treatment processes, mechanical aerators and diffusers happen to be the largest energy consumers.

Aeration mixes the liquor and is also responsible for delivering O2 to the microorganisms present in the tank. However, "Dead zones" can develop; giving rise to anaerobic/anoxic conditions in inadequately mixed areas. Minimizing these areas is favorable as it can decrease sludge bulking, which diminishes the settling efficiency in secondary clarifiers and also can reduce unwanted odors.

1.4.3 Secondary Clarifiers

Gravity settling is used to remove biomass and other solids moving out of the aeration tank through clarifiers. The flow rate of the liquid is regulated so that the settling velocity of the biosolids existing in the liquid is greater than the up-flow velocity of the liquid. To control the required biomass and prolong the solids' contact time with the wastes some of the settled biosolids are sent back to the aeration tank.

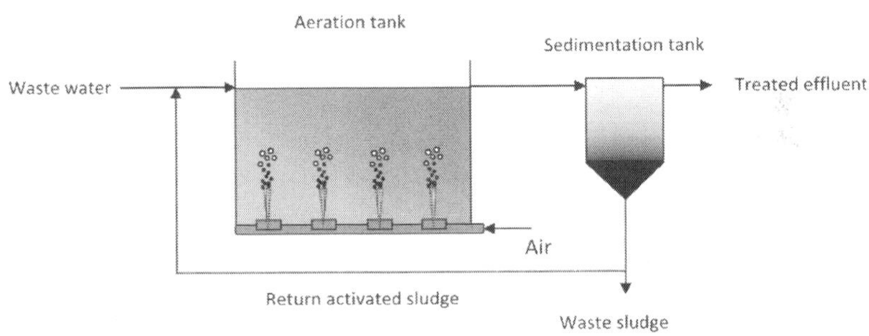

Figure 1.2: Activated sludge process.

[Source: https://www.researchgate.net/publication/234167418_Wastewater_treatment_Biological]

1.4.4 Essential Operating Criteria in Activated Sludge Systems

Table 1 depicts the major functioning parameters and standard values for activated sludge systems. The guidelines are used to pseudo-control and direct biosolids levels, they deeply influence process performance. Mixed-liquor suspended solids (MLSS) are the entire suspended solids in the aeration tank; which describes the number of solids in a specific volume of the water (milligram of solids per liter). The real biomass fraction of the solids is predicted as the solids that can be volatilized at 550°C. Mixed-liquor volatile suspended solids (MLVSS) are the volatile fraction. Thus, for active biomass treating waste MLVSS is often used as a proxy. In many activated sludge systems, MLVSS varies from 70% to 90% of the MLSS concentration (Qasim, 2017).

1.4.5 Solid Retention Time

The sludge age or solids retention time (SRT) is the average cell residence time of cells in the reactor; this happens to be the most crucial design parameter, in activated sludge systems. By controlling the pace at which surplus sludge is wasted and is affected by the state of the hydraulic flow through the reactor the SRT can be administered. It is the ratio of all solids leaving the system and the total solids in the system.

$$\text{SRT} = VX/(QXe + QwXw)$$

where SRT is the average cell residence time (day); X is the mixed liquor suspended solids concentration (mg/L); Xw is the concentration of waste sludge suspended solids (mg/L); X e isthe effluent suspended solids concentration (mg/L); Q is the volumetric flow rate (e.g., L/day); Qw is the waste sludge flow rate (e.g., L/day); and V is the volume of aeration basin (e.g., L);.

1.4.6 Sludge Volume Index

Used for defining the settling qualities of the sludge, the sludge volume index (SVI) is another basic parameter. The SVI is described as a secured volume of 1g of sludge (mL/g) after a 30 min settling period. An evident separation between the water and the sludge can be viewed in a well-settled sludge. In case of problems like ashing, pinpoint floc formation of tiny, poorly settling floc, or bulking the junction between the water and the sludge may not be viewed clearly. Issues in the aeration tank can lead so these scenarios. They are the reason behind the decrease in effluent quality due to poor settling in the clarifier (Pol &Lettinga, 1986).

1.4.7 Quantity of Dissolved Oxygen

In an activated sludge system, microbes need sufficient oxygen to oxidize organics present in the waste. The fundamental oxidation reaction for organics deterioration can be :

The basic oxidation reaction for organics degradation can be calculated as (stoichiometry not provided)

CHON + O2 + microorganisms ® CO2 + H2O + NH3 + more microorganisms

Organics are utilized by microorganisms, and new microbial cells are produced with a ratio of organisms created to the organics being utilized in the sludge yield.

Table 1.1: Standard ASP design parameters

Process components or variables	Aeration tank Depth (m) Width (m)	SRT (day)	MLSS (kg/m^3)	SVI (kg/m^3)	F/M	Organic loading rate (kg COD/ m^3day)	Oxygen require-ment (kg/ kg COD removed)
Typicalval-ues	5–8 7–12	5–15	1500–4000	40–150	0.2–0.4	20–60	1.4–1.6
Reference	[2]	[2]	[9]	[7]	[6]	[6]	[9]

Air pumps are more widely used to aerate the wastewater because of their lower operational and maintenance costs (McCarty & Smith, 1986; Noykova et al., 2002).

Diffusers and mechanical aerators are used to deliver oxygen to an aeration tank. Even though the procedure is carried out determines the amount of oxygen needed for the system, the design aims to reduce the addition of oxygen because of energy expenses. By controlling the velocity of the air pipes or the pace of the air pump, the quantity of the dissolved oxygen can be regulated. As air pumps have lowered functional and maintenance costs, they are increasingly being used to aerate wastewater.

1.4.8 Food-to-microorganism Ratio

Described as the quantity of sustainable organic material [milligrams of 5-day biological oxygen demand (BOD5)] feasible for the number of existing microbes (mg MLVSS) per day, food-to-microorganism ratio (F/M) is an appropriate measure for controlling and planning the aeration tank's functions (Bécares Mantecón, 2006).

$$F/M = (QSo)/X$$

Whereas, *So* is the influent BOD5 concentration (mg/L);*Q* is the volumetric flow rate (L/day); *X* is the MLVSS concentration (mg/L) and F/M is the food-to-microorganism ratio (day-1).

Differences in the F/M ratio set for all treatment systems are based on the system's framework, and values can vary remarkably. To preserve an F/M ratio that is optimal for the particular activated sludge design, the MLVSS is mostly adjusted by altering the secondary clarifier's return activated sludge rate as the influent BOD cannot be regulated (Lee et al., 2001).

1.4.9 Organic Loading Rate

The chemical oxygen demand (COD), total organic carbon content, or BOD5is typically used to calculate the quantity of organic matter in wastewater. Unfinished treatment is the result of the absence of sufficient organisms in the aeration tank or overabundance of organics in the influent (Dortch, 1996; Kim et al., 2006).

1.4.10 Standard Microorganisms in Activated Sludge Systems

A diverse and complicated mixture of microorganisms, makeup the activated sludge (Henze et al., 2008). The most important classes are algae, protozoa (e.g., flagellates, ciliates, and rotifers), bacteria, viruses, and fungi. The basic reason for post-biological disinfection processes to be carried out in treatment plants is the presence of pathogenic bacteria and viruses in wastewater.

1.5 ATTACHED GROWTH PROCEDURES

Just like activated sludge systems, attached growth processes like trickling filters (Fig. 1.3) are capable of attaining similar treatment goals. In these systems, the conversion procedures are mostly mass transport limited: major removal of the substrate is achieved by the microbes in the outer covering of the biofilm (Rouse et al., 2004; Loupasaki&Diamadopoulos, 2013). Despite the water flowing down the filter and the growth of biofilm, the material that makes up the foundation of trickling filters is selected, based on providing adequate and abundant pore space to let air through it. In addition to being used for the oxidation of ammonia and carbon, trickling filters are also capable of denitrification, after the optimization of the system; through the convection of air (Zehnder, 1988; Colleran et al., 1995).

1.6 ANAEROBIC WASTEWATER TREATMENT PROCESSES

Various industries mostly pursue anaerobic treatment technologies, based on their suitability and need. The summary of the various drawbacks and benefits of this procedure, while treating different wastes has been shown in Table1.2. The syntrophic and consequential metabolic interactions of different trophic units of prokaryotes, which include sulfate-reducing bacteria (SRB), acetogens, methanogens, and fermenters, allow further decomposition of organic matter under anaerobic circumstances. Thus, simple compounds like, ammonia, carbon dioxide, sulfide, and methane are formed from complex organic compounds, as a result of these metabolic interactions among these microbial groups (Polprasert et al., 1994; Braun, 2007). Requiring various microorganisms in every stage, the digestion phase is completed in four main reaction processes (Speece, 1983; Bitton, 2014).

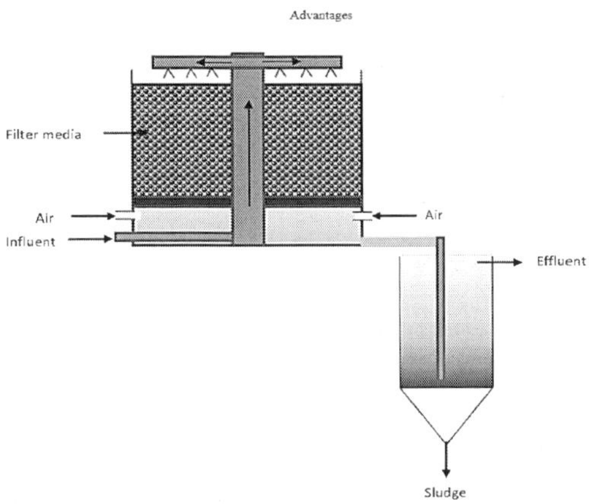

Figure 1.3: Trickling filter for aerobic respiration.

[Source: https://www.researchgate.net/publication/234167418_Wastewater_treatment_Biological]

- *Stage 1:* *Hydrolysis*—proteins, lipids, and carbohydrates are predominant in making up the material in organic waste. The action of these microorganisms and their extracellular enzymes break down the large and complicated matter into simpler compounds. Soluble monomers like glycerol, glucose, fatty acids, and amino acids are converted from complex organic molecules, such as lipids, lignins, cellulose, and proteins by these hydrolyzing organisms. *Lactobacillus*, *Bifidobacterium*,*Bacteroides*, and *Clostridium*are the hydrolytic microbes that are mainly responsible for the solubilization and hydrolysis. Fermentative acidogenic bacteria utilize these hydrolyzed products in the next stage (Zeikus,1980).
- *Stage 2:* *Acidogenesis*— Hydrogen, short-chain organic acids like valeric, propionic,succinic, isovaleric, lactic, butyric, formic, isobutyric, lactic and acetic acid; carbondioxide,ketones

(methanol, acetone, glycerol, and ethanol) and alcohols are modified from simple organic matter like long-chain fatty acids, sugars and amino acids by fermentative acidogenic bacteria. Normally, in an anaerobic digester, acidogenic bacteria are the most abundant bacteria and grow at a fast pace (Gujer&Zehnder, 1983).However, as suggested by the considerable activity of these organisms in the anaerobic digestion process, acidogenesis is by no means the rate-limiting step (Dolfing, 1988). The microorganisms used specifically for the acetogenesis stage, further refine the volatile acids made in this process.

- *Stage 3:* *Acetogenesis*—organic acids and alcohols are transformed into carbon dioxide, acetate, and hydrogen by acetogenic bacteria, which are also called obligate hydrogen-producing acetogens. The modified products are then utilized by SRB and methanogens. Methanogens and acetogenic bacteria have a strong symbiotic relationship. Acetogenic conversions demand a low hydrogen pressure environment, which is obtained when hydrogen is utilized by SRB and methanogens (Vavilin et al., 2008).
- *Stage 4:* *Methanogenesis*—this stage is mostly considered as the rate-limiting step of the anaerobic process, due to the reduction in methanogen's growth rate, however, some cases also present hydrolysis as the rate-limiting step (Seghezzo et al., 1998). The conversion of methanogenic archaea into hydrogen, methanol, acetate, formate, and methylamine formed in the previous stages takes place in the final process.

1.6.1 UASB REACTORS

As depicted in Fig. 1.4, the UASB reactor has an attached self-immobilized cell structure that contains an upper liquid layer and a lower layer of packed sludge bed (sludge blanket). It happens to be the most frequently and extensively used anaerobic reactor (Schink, 1987; Kato et al., 1994).

Clusters of bacteria present in the form of a floating blanket change complicated organic matter to hydrogen, methane, and carbon dioxide; when wastewater moves upwards through the sludge bed (Ahammad et al., 2011).

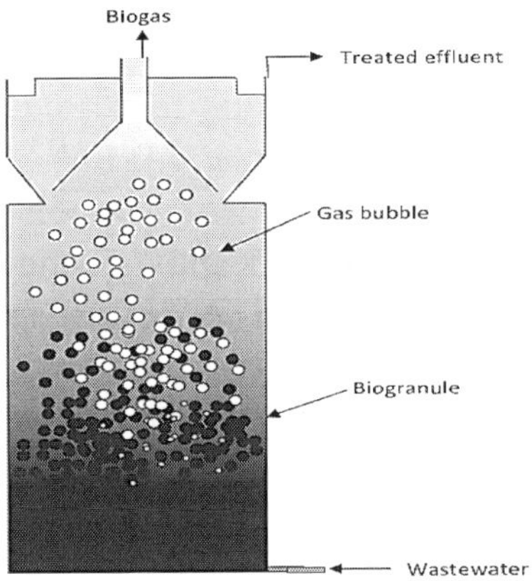

Figure 1.4: Up-flow anaerobic sludge blanket reactor.

[Source: https://www.researchgate.net/publication/234167418_Wastewater_treatment_Biological]

The granular sludge (1–5 mm in diameter) is characterized by good settling properties, specific activity, and high biomass content (MLVSS). While the effluent recirculation causes the upward flow of the liquid inside the reactor; whichcan sustain a high SRT, due to the presence of a high density of biomass in the self-immobilized granular sludge that does not demand support material and is distinct from the hydraulic retention time (HRT). To obtain the appropriate level of biodegradation, there is a high need for HRT; and the samehappens to be a major setback of UASB. Extensive reactor volume requires high maintenance of HRT. These issues can be controlled by utilizing HAR, where UASB operation is linked with the benefits of AFBR by regulating a high up-flow velocity (4–8 m/hr) within the reactor (Dolfing, 1986). With increased up-flow velocity, improved mass transfer is achieved in the reactor, demonstrating higher regression with decreased HRT operation. The major objective of these reactors is to obtain superior degeneration of waste and boost the formation of biogas (methane) mainly in an anaerobic reactor with reduced capacity (Reyes-Avila et al., 2004; Zhu et al., 2008).

1.6.2 Essential Operation Guidelines in Anaerobic Reactors

Despite regulated operational conditions, various operating parameters, for example, HRT, pH, nutrients, and temperature, among others; and their disruptions become apparent at times of industrial wastewaters treatment in

1.7 BIOLOGICAL REMOVAL OF NITROGEN

Typical removal of biological nitrogen is carried out in two steps, firstly nitrification and then denitrification; however, it should also be noted that due to low production and microbial activity, the process takes longer time than previously hypothesized Accurately controlled aerobic conditions are demanded by nitrification, which comprises chemolithoautotrophic oxidation of ammonia to nitrate. Two consecutive phases of oxidation lead to the oxidation of ammonia to nitrite (ammonia oxidation) and from nitrite to nitrate (nitrite oxidation).Thus, it should be noted that *Nitrosolobus, Nitrosopira,Nitrosomonas*, *Nitrosovibrio,* and *Nitrosococcus* are the nitrifier microorganisms that are used to oxidize ammonia to nitrite. While *Nitrocystis, Nitrospira, Nitrococcus,Nitrobacter,* and *Nitrospina* are engaged in the production of nitrate, during the nitrite oxidation phase (Jetten et al., 2001; 2002). Moreover, carbon dioxide is used as a carbon source, while oxygen can acts as an electron acceptor, by the various microorganisms involved in these steps. Influent ammonium concentration, reactor configuration, and substrate type determine the ammonia uptake rate. Denitrification is a heterotrophic bioconversion procedure, performed in anoxic conditions by the heterotrophic denitrifiers. It is the second phase of the nitrogen removal process. Normally, *Halobacterium, Paracoccus,Pseudomonas,Thiobacillus*, and *Alcaligenes* are present in denitrification systems (Hellinga et al., 1998). Denitrifiers yield nitrogen gas by oxidizing nitrogen compounds (NO2– and NO3–). These denitrifiers utilize used nitrite and/or nitrate as organic matter, such as carbon and energy source and terminal electron acceptors.

Based on the attributes of wastewater, some elaborate procedures such as anaerobic ammonium oxidation (Anammox), autotrophic nitrogen removal (Canon), and partial nitrification are being exercised in various treatment plants. As compared to the traditional two-phase process, a collective approach, involving anammox and partial nitrification is beneficial; as only a small quantity of sludge is produced, a lower amount of oxygen and energy is needed along with no additional requirement of carbon (Jokela et al., 2002).

1.7.1 The Sharon Process

Elimination of ammonia through nitrite production is achieved through the Sharon (single-reactor high-activity ammonium removal over nitrite) process (Van Loosdrecht, 2004). This procedure is conducted in a single reactor with periodic aeration and supports both heterotrophic and autotrophic denitrification. The addition of menthol is utilized in the Sharon process as a source of carbon. This process is appropriate for eradicating nitrogen from waste streams possessing high ammonia concentrations (>0.5 g/L), however, it is inappropriate for all wastewaters; because of the high-temperature dependency.

1.7.2 Anaerobic Oxidation of Ammonium

Anaerobic ammonium oxidation (Anammox) is a process, in which, ammonia is transformed into nitrogen, by the action of a group of planctomycete bacteria; it is an extremely lithoautotrophic, exergonic conversion process (Van Loosdrecht, 2004). By utilizing nitrite as the electron acceptor in an anoxic environment, these microbes possess the ability to oxidize ammonia to gaseous nitrogen; they use CO2 as the only source of carbon.

1.7.3 Collective Removal of Nitrogen

Ammonia needs to be preoxidized to nitrite before entering the anammox process. Therefore, nitrite is required as a precursor for the anammox treatment of ammonia-rich wastewater. To enhance the nitrogen removal efficiency, a partial Sharon process may be carried out before the anammox procedure. The Sharon process accomplishes partial nitritation (conversion of 55%–60% of ammonium to nitrite) without heterotrophic denitrification. The Anammox reactor then administers nitrite-rich waste. While in the partial Sharon–Anammox digester, a total of 83% of ammonia nitrogen elimination can be achieved from the waste stream, which has a total nitrogen load of 0.8 kg N/m3/day (Barnard, 1975).

1.7.4 Canon Process

The Canon (completely autotrophic nitrogen removal over nitrite) process consists of two classes of aerobic and anaerobic microorganisms (e.g., *Nitrosomonas*and planctomycetes) that execute two successive reactions in a sole and aerated reactor. It is a mixture of the anammox and partial nitritation processes. The nitrifiers produce nitrite by oxidizing ammonia and utilizing oxygen. The anammox process requires an anoxic condition which is formed by the consumption of oxygen. Operational specifications like temperature, dissolved oxygen, nitrogen-surface load, and biofilm thickness are important for determining the performance of the Canon process (Mino et al., 1998).

Lettinga, G. (1998). A review: the anaerobic treatment of sewage in UASB and EGSB reactors. *Bioresource technology*, *65*(3), 175-190.

69. Seghezzo, L., Zeeman, G., van Lier, J. B., Hamelers, H. V. M., & Lettinga, G. (1998). A review: the anaerobic treatment of sewage in UASB and EGSB reactors. *Bioresource technology*, *65*(3), 175-190.
70. Servais, P., Billen, G., & Bouillot, P. (1994). Biological colonization of granular activated carbon filters in drinking-water treatment. *Journal of Environmental Engineering*, *120*(4), 888-899.
71. Shipin, O. V., Lee, S. H., Chiemchaisri, C., Wiwattanakom, W., Ghosh, G. C., Anceno, A. J., & Stevens, W. F. (2007). Piggery wastewater treatment in a tropical climate: biological and chemical treatment options. *Environmental technology*, *28*(3), 329-337.
72. Show, K. Y., Lee, D. J., & Pan, X. (2013). Simultaneous biological removal of nitrogen–sulfur–carbon: recent advances and challenges. *Biotechnology advances*, *31*(4), 409-420.
73. Speece, R. E. (1983). Anaerobic biotechnology for industrial wastewater treatment. *Environmental science & technology*, *17*(9), 416A-427A.
74. Stouthamer, A. H. (1992). Metabolic pathways inParacoccus denitrificans and closely related bacteria in relation to the phylogeny of prokaryotes. *Antonie van Leeuwenhoek*, *61*(1), 1-33.
75. Tchobanoglous, G. (2003). Fundamentals of biological treatment. *Wastewater engineering: Treatment and reuse*, 1, 611-635.
76. Teske, A., Alm, E., Regan, J. M., Toze, S., Rittmann, B. E., & Stahl, D. A. (1994). Evolutionary relationships among ammonia-and nitrite-oxidizing bacteria. *Journal of bacteriology*, *176*(21), 6623-6630.
77. Van den Heuvel, J. C., Beeftink, H. H., & Verschuren, P. G. (1988). Inhibition of the acidogenic dissimilation of glucose in anaerobic continuous cultures by free butyric acid. *Applied microbiology and biotechnology*, *29*(1), 89-94.
78. Van Loosdrecht, M. C. M. (2004). Recent development on biological wastewater nitrogen removal technologies. In *Proceedings of the international conference on wastewater treatment for nutrient removal and reuse (ICWNR'04). Pathumthani, Thailand, January* (Vol.1, pp. 26-29).
79. Vavilin, V. A., Fernandez, B., Palatsi, J., & Flotats, X. (2008). Hydrolysis kinetics in anaerobic degradation of particulate organic material: an overview. *Waste management*, *28*(6), 939-951.

80. Zehnder, A. J. (1988). *Biology of anaerobic microorganisms (Vol.1, pp.1-22)*. John Wiley and Sons Inc.
81. Zeikus, J. G. (1980). Microbial populations in digesters. In *Anaerobic digestion:[proceedings of the first International Symposium on Anaerobic Digestion, held at University College, Cardiff, Wales, September 1979]/edited by DA Stafford, BI Wheatley and DE Hughes (Vol.1, pp.1-33)*. London, Applied Science Publishers,[1980]..
82. Zhou, P., Su, C., Li, B., & Qian, Y. (2006). Treatment of high-strength pharmaceutical wastewater and removal of antibiotics in anaerobic and aerobic biological treatment processes. *Journal of Environmental Engineering, 132*(1), 129-136.
83. Zhu, G., Peng, Y., Li, B., Guo, J., Yang, Q., & Wang, S. (2008). Biological removal of nitrogen from wastewater. In *Reviews of environmental contamination and toxicology* (Vol.1, pp. 159-195). Springer, New York, NY.

CHAPTER 2

MICROBIOLOGY AND ECOLOGY OF WASTEWATER TREATMENT

CONTENTS

2.1 INTRODUCTION

As the name indicates, biological wastewater treatment takes place solely by biological processes. The biological procedures reproduce, in a specific way. Within water bodies, by natural mechanisms organic matter converts in the mineralized (inert) products, representing the *self-purification* phenomenon. The same phenomenon takes place in a wastewater treatment plant, but technology is introduced in it(Rowan et al., 2003; Daims et al., 2006; Narihiro & Sekiguchi, 2007). The main objective of this technology is to develop the process of *purification* at **higher rates** (compact solutions) and under **controlled conditions** (Arceivala, 1981; Abnt, 1989; Boon et al., 2002).

Understanding the sewage treatment's microbiology is elementary for developing the design and for operating biological treatment systems. Based on empirical criteria, in the past the functioning of the treatments were designed by the engineers.

Over recent decades, recognition of Sanitary and Environmental Engineering as multidiscipline took place, and to understand this process, biologists have fundamentally contributed (Da Silva et al., 2006; Ferreira et al., 2007). Decreasing the empiricism's level and expanding rational knowledge has permitted the systems to be operated and designed with a comparatively more solid base. The outcome has surely reduced costs and increased efficiency (Benefield &Randall, 1980; Asce, 1990; Amann et al., 1998).

2.2 PRESENCE OF MICROORGANISMS IN BOTH: WATER AND WASTEWATER

The branch of science (biology) that deals with microorganisms is called *microbiology*. Microorganisms play a key role in the matter of water quality, because in certain environments they are predominant;as they are associated with diseases such as waterborne diseases, and as they play role in the purification process of wastewater. We can observe microorganisms through a microscope only (Boon,1980; Sharma, 2007; Akpor et al., 2014).

Certain groups of microorganisms have characteristics of plants, while other groups have characteristics of animals. In the past, living organisms were divided into two large kingdoms only, **Animals** and **Plants**; whereas microorganisms were found in both of these subdivisions (Zhao et al., 2006; Poleto et al., 2016).

Eventually, with time, biologists adopted a better division; it was a more practical one as microorganisms were placed in the separate two kingdoms, the **Protists** (it contains simple creatures, but having a separate nucleus, e.g. protozoa, fungi, and algae) and the **Monera** (more simple creatures, not having a separate nucleus, e.g. cyanobacteria, archaea, and bacteria).

The major difference between the microorganisms (the protists and the monera) and other organisms (including animals and plants) is cellular differentiation, a high level of it is found in animals and plants. In protist and monera, the individual's cells are functionally and morphologically similar; which further reduces the development capacity and its adaptation. However, a functional division is observed in the organisms having cellular differentiation. Tissues are found in higher organisms. These tissues are formed by the combination of the same type of differentiated cells in smaller or larger groups; and are responsible for constituting the organs, such as lungs that are further combined to form systems like vascular systems, respiratory systems etc. . The species' developmental level is indicated by the cellular differentiation level (Sobsey,1989; Olaolu et al., 2014).

The living world's general characteristics of the kingdoms are presented in Table2.1, and the primary characteristics of groups comprising the kingdom of the protist and monera are listed in Table 2.2.

Table 2.1: Living world's general characteristics of the kingdoms

Characteristics	**Monera and Protists**	**Plants**	**Animals**
Cell	Multicellular or unicellular	Multicellular	Multicellular
Cellular differentiation	Non-existent	High	High
Energy source	Organic matter, light	Light	Organic matter
	or inorganic matter		
Chlorophyll	Present or absent	Present	Absent
Movement	Mobile or immobile	Immobile	Mobile
Cell wall	Present or absent	Present	Absent

Table 2.2: Primary characteristics of microorganisms (protists and monera) (La Rivie´re, 1980; Kubo et al., 2001)

	Monera(prokaryotes)		**Protists(eukaryotes)**		
Characteristic	**Bacteria**	**Cyanobacteria**	**Algae**	**Protozoa**	**Fungi**
Nuclear membrane	Absent	Absent	Present	Present	Present

Photosynthesis	Minority	Majority	Yes	No	No
Movement	Some	Some	Some	Mobile	Immo-bile

2.3 BIOLOGICAL CELLS

In general, most of the living cells are similar to each other. A brief explanation of their leading components is given below (LaRivie´re,1980;Tchobanoglous&Schroeder, 1985).

As an outer boundary, the cells have a *cell membrane*. It is flexible and serves its function as a barrier or better called selective barrier. This barrier is between the external environment and the things contained by the cell. It plays a primary role in choosing and selecting the substance to enter or leave the cell; hence cell membrane is semi-permeable (Mourant et al., 1998; Yi et al., 2006; Kang et al., 2008). However, several organisms such as plants, fungi, algae, and bacteria have another external layer named *cell wall*. It is constituted by a rigid material, which gives structural protection to the cell; it also offers protection against osmotic alterations and mechanical impacts (White, 1976, Wilson, 1981; Wpcf & Asce, 1988). The cell wall is not considered semi-permeable; hence it does not play a role in regulating the consumption of substances (dissolved ones) in the neighboring medium. Some bacteria have another external layer of a gelatinous material, around their cell wall; which is named *gelatinous layer* or mostly *capsule*. In the condition, when motility is present in individual cells; they mostly have *cilia* or flagella (Deeley, 1955; Glazier & Graner, 1993).

In the interior, cells composed of a colloidal suspension of carbohydrates, proteins, and other complex structures of organic matter. They also contain *organelles*. All these combine to constitute the *cytoplasm* (Wef & Asce, 1992; Wanner,1994).

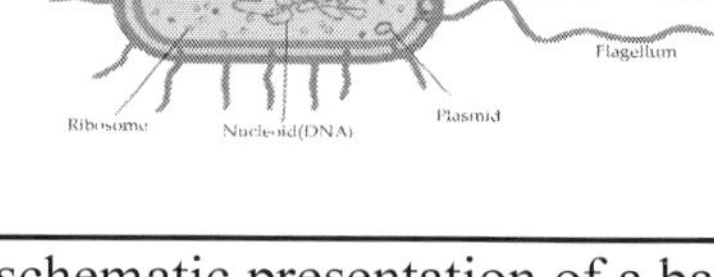

Figure 2.1: Simple and schematic presentation of a bacterial cell.

[Source: http://classnotes123.com/bacteria-microorganism-friend-and-foe-class-8-cbse-notes/]

A genetic material beneficial for reproduction named *nucleic acid* is present in each cell. Ribonucleic acid (RNA) is of key importance for protein synthesis and it is present in ribosomes that are found in the cytoplasm (Fulwyler, 1965; Höfling & Franosch, 2013). A *nuclear area* is contained by the prokaryotic cells (bacterial cells), on the other hand, a membrane-enclosed *nucleus* is found in eukaryotic cells. In this nuclear area or nucleus, deoxyribonucleic acid (DNA) is present. DNA contains all the necessary genetic information, needed by all the cell components for reproduction. The prokaryotic cell's cytoplasm contains DNA in the shape of small structures named plasmids. Figure 2.1 represents a simple and schematic presentation of the bacterial cell (Brownstein Tarr, 1979; Sohn et al., 2000).

2.4 CARBON AND ENERGY SOURCES FOR MICROBIAL CELLS

Living organisms require (a) energy (b) nutrients (potassium, phosphorus, magnesium, nitrogen, calcium, sulfur, etc.) and (c) carbon for the functions of reproduction, locomotion, growth, and others.

There are two elementary organism types, referring to the *carbon source* (Tsuge, 2002; Catal et al., 2008):

1. **Autotrophic organisms**. *Carbon dioxide* (CO2) as a carbon source;
2. **Heterotrophic organisms**. *Organic matter* as a carbon source.

Referring to the *energy source*, there are two fundamental organism types:

1. **Phototrophic organisms**. *Light* as an energy source
2. **Chemotrophic organisms**. The *energy coming from chemical reactions* as an energy source.

Table 2.3 shows the combination of these four types. Except for maturation and facultative ponds, in the majority of sewage treatment processes, significant penetration of light is not observed in the biological reactors containing a liquid; it is due to liquid's high turbidity. This extremely limits the presence of microorganisms using light as their source of energy (photoheterotrophs and photoautotrophs). Hence, in this case, the really important organisms are the *chemoheterotrophs* (the ones that are responsible for the reactions that take place in biological treatment) and the *chemoautotrophs* (the ones responsible for nitrification). To keep this simple, the later ones will be called only *heterotrophs (Kim et al., 2000; Fei et al., 2011)*.

Table 2.3: Classification of microorganisms formed on sources of carbon and energy (Tchobanoglous & Schroeder, 1985; Metcalf & Eddy, 1991)

Classification	Energy source	Carbon source	Representative organisms
Photoautotrophs	Light	CO_2	Algae, higher plants,
			Photosynthetic
			Bacteria
Photohetero-trophs	Light	Organic matter	Photosynthetic
			Bacteria
Chemoauto-trophs	Inorganic matter	CO_2	Bacteria
Chemohetero-trophs	Organic matter	Organic matter	Bacteria, protozoa, fungi, animals

2.4.1 Metabolism of Microorganisms

All the chemical processes, occurring simultaneously in the cell are termed as *metabolism*. We can divide metabolism into two categories (LaRivie´re,1980):

1. ***Catabolism* or *Dissimilation***: Energy production reactions in which decomposition of substrates takes place;
2. ***Anabolism* or *Assimilation***: Utilizing the energy that was released in the process of dissimilation, cellular material for growth is formed by these reactions.

The growth and reproduction process of organisms takes place at the cost of energy that can be released in the process of dissimilation. In the dissimilation process, energy in the organic compounds, stored in the form of chemical, releases and gets changed in the cellular material's assimilation process; this balance between the catabolism (negative) and the anabolism (positive) results in the net growth.

In each of the categories, the chemical transformation takes place in the order of intricate and diverse intermediate reactions; a specific kind of enzyme catalyzes each of them. Mostly, enzymes are situated inside the cell; such enzymes are termed as "*endoenzymes*or" i.e. intracellular enzymes. Some enzymes are released in the outer medium (external) and are termed as "*exoenzymes*" or extracellular enzymes. They are important because they are responsible for hydrolysis reactions, taking place outside of the cell; and are responsible for converting complex and large substrate molecules in simple and small molecules, the ones that can pass the cell membrane and make themselves available for cellular consumption (Barbosa et al., 2001; Vinuselvi et al., 2012).

Table 2.4: Major characteristics of the fermentative and oxidative catabolism

Characteristic	**Oxidative catabolism (respiration)**	**Fermentative catabolism (fermentation)**
Electron donor	Organic matter	Oxidized organic matter
Electron acceptor	External: inorganic compound (sulfate, nitrate or oxygen)	Internal: reduced organic matter
Final products resulting from the organic matter	One (CO_2)	Minimum two (CO_2 and CH_4)
Carbon's forming the finalproduct	Oxidized inorganic carbon (CO_2)	Oxidizedinorganic carbon (CO_2) + Reduced organic carbon (CH_4)
Carbon's oxidation statein finalproduct	4 +(CO_2)	4 + (CO_2) and 4 - (CH_4)

Organic matters are removed from the sewage by the process of catabolism, also known asdissimilation. Catabolism has two types of treatments, one is *fermentative catabolism* (fermentation of organic matter) and other is *oxidative catabolism* (oxidation of organic matter) (vanHaandelandLettinga, 1994; Lettinga, 1995; Lettinga et al., 1996):

1. ***Fermentative catabolism***: In this type, there are no oxidants. The process takes place because of the electrons rearrangement in the fermented molecule. Electrons are rearranged in a way, so that two or more than two products are formed. Generally, various sequences of fermentation are required to stabilize the product,; important for it to be susceptible for longer time duration. ***Oxidative catabolism: They*** are the redox reactions; in them, an oxidizing agent (nitrate, sulfate, or oxygen) is present for the oxidization of organic matters (VonSperling,1983a; 1983b; Viesmann& Hammer, 1985).

The basic characteristics of the fermentative and oxidative catabolism are represented in Table 2.4.

2.4.2 Energy *Generation in Cells* (Microbial)

In microbial cells, the accomplishment of energy generation depends on microorganisms; if these microorganisms adopt a fermentative catabolism, then the process of oxidation is known as fermentative catabolism, whereas oxidative catabolism is known as the respiration.

The process of respiration is not limited to the processes, which are involving the consumption of oxygen. Generally, oxidation refers to the loss of electrons (one or more). This electron loss takes place from the oxidized substance (in the process of oxidation, negative charges are given by the substance as electrons, when they are going to ahigher oxidation state. The oxidized substances might be inorganic compounds (reduced) or organic matter; hence both of them can be **electron donors**. The electrons from these molecules (oxidized ones) are transferred by the means of enzymes, through biochemical reactions to some oxidizing agents, which are themselves inorganic compounds; and receive the denomination of **electron acceptors.** Further resulting in reduced oxidation state of electron acceptors (Badziong & Thauer, 1978; Thurnheer et al., 1986).

Table 2.5: Electron acceptors usually used in the respiration (oxidation reactions) process in sewage treatment (energy release is listed in decreasing order)

Conditions	Electron acceptor	Form of the acceptor after the reaction	Process
Aerobic	*Oxygen* (O_2)	H_2O	Aerobic metabolism
Anoxic	*Nitrate* (NO_3^-)	Nitrogen gas (N_2)	Nitrate reduction
			(denitrification)
Anaerobic	*Sulfate* (SO_4^{2-})	Sulfide (H_2S)	Sulfate reduction
	Carbon dioxide (CO_2)	Methane (CH_4)	Methanogenesis

Table 2.5 lists the electron acceptors usually used in the respiration process (energy released is listed in decreasing order) (Von Sperling, 1990; 1993; 1994; 1997).

In the medium, when a variety of electron acceptors are present, the one producing the highest amount of energy is utilized by the microorganisms. Due to this reason, at first, the *oxygen (dissolved)* is utilized, when it exhausts; the system stops being **aerobic**. If in liquid medium *nitrate* is available, then the organisms start using it in their respiration process; further carrying out the process of denitrification (conversion of nitrogen gas from nitrates).

These conditions are termed as **anoxic** (nitrates are present but dissolved oxygen is absent). Occurrences of strict anaerobic conditions are observed, when nitrates are finished. In these conditions, *carbon dioxide* converts into methane and *sulfate* reduce to sulfides. Some of the substances are not utilized, while some substances release great energy (Arceivala,1981; Von Sperling, 1999; 2002).

In two pathways, the methanogenesis stage might take place. In the first one, *hydrogenotrophic methanogenesis* oxidative process occurs, in which methane is produced from hydrogen; in this process carbon dioxide plays a role of an electron acceptor, and accordingly reduces itself to methane (VonSperling&Fro´Es,1999). Concerning global conversion, this pathway is of less importance, but all methanogenic organisms can practically make it. *Acetotrophic methanogenesis* (acetate production from methane) is the second pathway. In this process, acetate is the organic compound that is converted into methane. Mostly the conversions take place through this pathway, though few bacteria species accomplish it (Lubberding,1995; Choi et al., 2004).

Some organisms functionally accommodate a variety of respiration conditions. The basic ones are (Marais &Ekama, 1976):

1. **Strict aerobic organisms**: In the respiration process, they use *free oxygen* only
2. **Facultative organisms**: Uses *nitrate* or preferably *free oxygen* as an electron acceptor
3. **Strict anaerobic organisms**: *carbon dioxide* or *suphates* is used as an electron acceptor and energy through respiration (aerobic) cannot be obtained.

More energy is released by aerobic reactions compared to the energy released from anaerobic reactions, owing to this fact, compared to anaerobic, the aerobic organisms stabilize the organic matter quicker and they are capable of reproducing themselves. As aerobic organism's reproduction rate is greater; the generation of sludge is also greater.

The energy generation reactions that mainly take place in anoxic, anaerobic, and aerobic conditions are:

Aerobic conditions:

$C6H12O6+6\ O2 \quad >6\ CO2+\ 6H2O$

Anoxic conditions: The reduction of nitrate(denitrification)

$2\ NO3-+2\ H+ ------------> N2+\ 2{,}5\ O2+\ H2O$

Anaerobic conditions: *Reduction of sulphate*

CH3COOH + SO42− + 2 H+------------->H2S + 2 H2O + 2 CO2

Anaerobic conditions: The reduction of CO2(hydrogenotrophic methanogenesis)

4 H2+ CO2 ------------->CH4+ 2 H2O

Anaerobic conditions: acetotrophic methanogenesis

CH3COOH ------------->CH4+ CO2

The organic matter's decomposition routes in the availability of electron acceptors are illustrated in Figure 2.2.

The transformation sequence that takes place in the treatment of sewage is a function of the compound's oxidation state and an electron acceptor; and is further measured by the oxidation-reduction potential (represented in millivolts). These reactions are illustrated in Figure 2.3.

The maximum amount of energy available is determined by the compound's oxidation state;while more amount of energy is contained by the the reduced compound. Energetic metabolism conserves as much energy as it can. The difference between the final product energy content (as given from the oxidation state) and its energetic content is the maximum energy available by the substrate's oxidation.

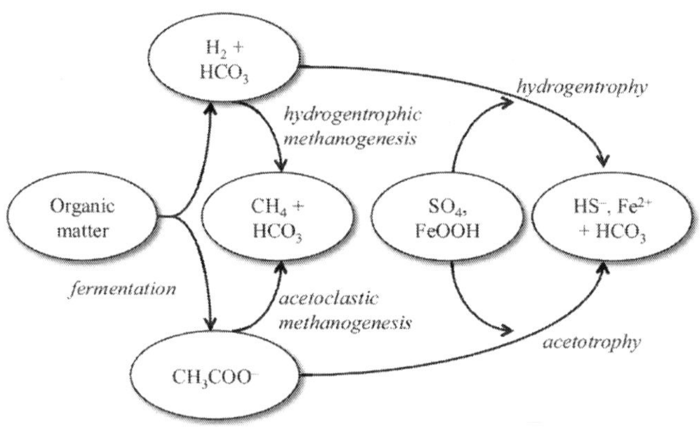

Figure 2.2: Organic matter's major routes of decomposition (Lubberding, 1995).

[Source:https://www.researchgate.net/publication/228477106_The_Thermodynamic_ladder_in_Geomicrobiology/figures?lo=1&utm_source=google&utm_medium=organic]

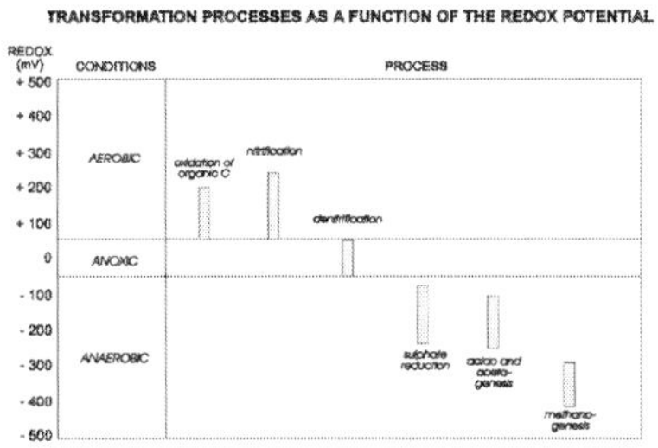

Figure 2.3: Sewage treatment's transformation sequence, as a function of the redox potential and the electron acceptor (Eckenfelder and Grau, 1992).

[Source: https://www.iwapublishing.com/sites/default/files/ebooks/9781780402093.pdf]

The following points apply (Eckenfelder, 1980; 1989):

1. The higher is the energy released, when the final product's oxidation state is greater. In CO2, carbon has a higher oxidation state. Hence, oxidation reactions, oxidizing the substrate carbon to CO2; employing aerobic respiration, releases more energy compared to fermentation reactions, such as ethanol.
2. A greater amount of energy is released when a substrate's oxidation state is lower. For example, in acetic acid's oxidation to CO2, less energy is released compared to the ethanol oxidation to CO2; since the oxidation state of carbon found in the acetic acid is higher than in ethanol.
3. CO2 never plays a role as an energy resource; this is because the carbon in CO2 cannot be oxidized as it is already at the highest oxidation state.

2.5 BIOLOGICAL WASTEWATER TREATMENT'S ECOLOGY

Role of microorganisms that they play in sewage treatment relies on the process that is being used. Concerning the facultative ponds, algae serve a basic role in producing oxygen through photosynthesis. Ponds are designed in a way that they optimize the algae present in the liquid medium and design helps in obtaining a balance between algae and bacteria. In the system of anaerobic treatment, the conditions are exclusive of developing microorganisms adapted (functionally) to oxygen absence or favorable conditions. In this case, methanogenic and acidogenic organisms are important (Daims et al., 2006; Lu et al., 2014).Mostly, protozoa and bacteria are consistedof the microbial lmass that is involved in the process (aerobic). The importance of other organisms (like rotifers and fungi) is lower but they might be found. While treating certain industrial wastewaters, fungi are important as they are capable of surviving in the reduced ranges of pH and also as they survive with little nitrogen. However, the sludge settle ability can be deteriorated by fungi having filamentousstructure; as a consequence efficiency of the process is reduced. Rotifers are effective when consuming organic matter (only small particles) and dispersed bacteria. The productive biologicalpurification process is indicated by the presence of rotifers in the effluent (Metcalf&Eddy,1991). It can be concluded that the diversity of species of various microorganisms is low in biomass (Wagner et al., 2002; Collins et al., 2006).

In the aerobic sewage treatment, the predominant microorganisms are represented in a sequence in Figure 2.4. In the microbial community, ecological interactions responsible for increasing the population of microorganisms are followed by the downturn of another group, in the light of selective attributes of the medium transforming. Soon after the sewage is introduced in the biological reactor, the left organic matter (BOD) achieves the maximum level. Protozoa of type amoeba might be found and the quantity of bacteria is still lower. All these are incompetent for the competition of available food, usually found at reactors startup (Werker et al., 2003; Hao et al., 2010).

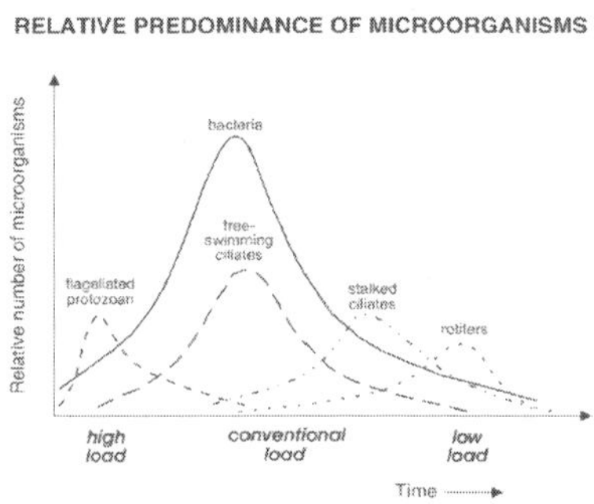

Figure 2.4: Relative predominance sequence of the microorganisms utilized in the sewage treatment (adapted from Ko¨nig, 1990; Metcalf & Eddy, 1991).

[Source: https://www.iwapublishing.com/sites/default/files/ebooks/9781780402093.pdf]

The growth in the population of bacteria is because of excessive substrate availability. Flagellated protozoa substitute amoebas. In the competition for the availability of food, flagellated protozoa are comparatively efficient because of their motility, these all are features of *high load* systems. With passing time along decreasing availability of organic matter, flagellated protozoa are substituted by the ciliated protozoa, as they survive on low food concentration, characterizing the *conventional load* systems operation. In this system, free-living ciliates are found in large numbers along with the low organic matter concentration (remaining BOD) and max amount of bacteria. Retention periods (long ones) characterizing the *low load* systems, rotifers, and ciliates consume the bacteria and the minimum amount of organic matter is left available (Ko¨nig, 1990; Kim & Aga,2007).

The two basic groups involved in the organic matter conversion are described below: bacteria (to which greater emphasis is given) and protozoa (Sakano et al., 2002; Gogina & Gulshin, 2014).

2.5.1 Bacteria

Bacteria are the microorganisms, in which the defined nucleus is absent i.e. Unicellular prokaryotic. They live in colonies or spend an isolated life. Table 2.6 lists the categories in which bacteria are classified based on their shape.

Table 2.6: Categories of bacteria based on their shape

Name	**Shape**	**Size**
Cocci (coccus, singular)	spheroid	0.5 to 3.0 μm in diameter
Bacilli (bacillus, singular)	rod	0.3 to 1.5 μm in width (or diameter)
		1.0 to 10.0 μm in length
Spirilla (spirillum, singular)	spiral	<50 μm in length
Vibrios	curved rod	0.6 to 1.0 μm in width (or diameter)
		2.0 to 6.0 μm in length
Various	filamentous	>100 μm in length

In bacteria, the flagella may or may not be present for locomotion and they have a cell wall that is less or more rigid. Their reproduction is mainly due to binary fission; besides sexual reproduction and spores formation (Branco,1976;Metcalf &Eddy,1991).

In the systems of wastewater treatment, the most important and the largest group is constituted by the bacteria. Considering the removal of BOD as a primary function in the treatment systems, the main agents are heterotrophic bacteria. Apart from being responsible for the conversion of organic matter, there exists a property in bacteria that they can agglomerate themselves in the form of structural units, having key implications in the treatment of wastewater. These structure units are biofilms, granules, and/or flocs (Catunda&Van Haandel, 1987; Chernicharo, 1997).

In addition to removing the carbonaceous organic matter, other objectives are also incorporated by the sewage treatment. These objectives rely on particular groups of bacteria. Hence, the phenomena given below can occur:

1. Ammonia to nitrite Conversion(a form of nitrification):chemoautotrophicbacteria
2. Nitrite to nitrate Conversion (a form of nitrification):chemoautotrophicbacteria
3. Nitrate to nitrogen gas Conversion (a form of denitrification): facultative chemo- heterotrophicbacteria

Section 1.3 represented the bacteria's cellular structure and Figure 2.1 illustrates it. The bacterial cell is composed of approximately 80% of water and the other 20% is dry matter. Around 10% of this dry matter is inorganic and 90% is organic. For the cell composition (approximated), the characterizations of formulas used widely are (Metcalf &Eddy,1991):

$C5H7O2N$ (the formula without phosphorus)

$C60H87O23N12P$ (the formula with phosphorus)

The ratio of C: H:O: N is same in either of two formulations. Key aspects needed to be mentioned are that these components must come by the medium and the unavailability of any of them would result in limited bacterial growth (in terms of population) (Harper & Pohland, 1986; Guo et al., 2013).

In the medium, the substrate available that is utilized by the bacteria is dependent on the particle size. In wastewater, mainly two fractions of the organic matter are found which are (a)*fraction that is easily biodegradable* and(b)*fraction that is slowly biodegradable*. Soluble form organic matter found in domestic sewage is easily degradable (Nielsen et al., 2009).

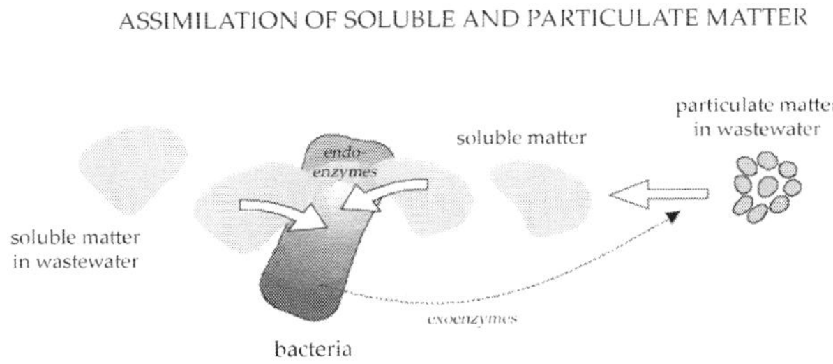

Figure 2.5: Assimilation mechanisms of converting particulate matter into the soluble matter and of soluble matter itself.

[Source: https://www.iwapublishing.com/sites/default/files/eb-ooks/9781780402093.pdf]

Owing to the soluble compound's small dimension, they are able to penetrate the bacterial cell by using cell membranes. With the help of *endoenzymes*, organic matter that is a soluble one is consumed inside the cell. Microorganisms having complex formulas (suspended organic matter or particulate) of compounds and large dimension; a transformation process that takes place outside the cell should be undergone by them, resulting in smaller molecules those can be assimilated. In *hydrolysis* reaction, with the help of *exoenzymes*, this action is completed. In this process, the consumption of energy is not accounted and electron acceptors are not used too. The end product of this reaction is presented in an easily biodegradable form; which penetrates inside the cell through, the cellular membrane, where it is consumed (IAWPRC, 1987) (see Figure 2.5).

The requirements of the environment may vary for the bacteria with the species. For instance, the type of bacteria called chemoautotrophic bacteria, involved in the process of nitrification is far more sensitive in terms of environmental conditions; compared to heterotrophic bacteria that stabilize carbonaceous organic matter. Generally, forbacterialgrowth, theoptimalgrowthrate takes place within limitedranges of pH and temperature, though it is possible for them to survive within a wider range.

There is a greater influence of temperature that lies beneath the optimallevel than to the temperature that lies above the level. Depending on the range of temperature, we can classify bacteria as *mesophilic,thermophilic* or *psychrophilic*. Table 2.7 shows these categories and their temperature ranges. In bacterial growth, another important factor is pH. Mostly, values of pH below4.0 or above9.5 are not supported by the bacteria, and around neutrality (that is almost 6.5 to 1.5) is the optimal value (Metcalf & Eddy, 1991; 2003).

2.5.2 Protozoa

Unicellular eukaryotic microorganisms having no cell walls are usually comprised of the protozoa groups. They do not have any cellular differentiation, though some have a bit complex structure along differentiatedregions inside the cell, these regions are responsible for performing different functions.

Table 2.7: Ranges of temperature for optimal bacterial growth (Metcalf & Eddy, 1991)

Type	*Temperature (°C)*	
	Range	**Ideal**
Psychrophilic	**10to30**	**12 to18**
Mesophilic	**20to50**	**25 to40**
Thermophilic	**35to75**	**55 to65**

A facultative heterotrophic or strictly aerobic organism represents the majority. In them, reproduction takes place bybinaryfission. Protozoa can feed on bacteria and are usually larger than them. Due to this, protozoa groups serve an important role in the food web. They allow the larger organisms to feed on bacteria indirectly; else it would be an inaccessible food form. Depending on motility mode and on a few structural characteristics, we can divide protozoa into various groups. The following are the ones of principal interest: flagellates, stalked, and free-swimming ciliates and amoebas (Branco, 1976;LaRivie´re,1980). Few of the species are pathogenic.

The following protozoa are important in the treatment of biological wastewater:

1. Organic matter consumption;
2. Free bacteria consumption;
3. Contribution in floc formation.

Regarding the contribution of forming flocs, the last aspect appears to be a less important mechanism (LaRivie´re,1977). The initial two aspects (organic matter consumption and free bacteria consumption) depends on the protozoa's feeding mode, its type varies. It is illustrated below (Horan,1990):

1. *Flagellates*. By active transport or by diffusion, it uses soluble organic matter. In terms of feeding mode, more efficient ones are the bacteria (De Korte &Smits, 1985).
2. *Ciliates and amoebas*. Around solid particles, the vacuole is formed (bacteria can be included) through the phagocytosis process. Inside the vacuole, enzymatic action takes place, and after it, particle's organic fraction is utilized.
3. *Ciliates* (principally). Predation of algae, bacteria, and other flagellated and ciliated protozoa.

Protozoa do contributeto removing organic matter from sewage but their major role in the treatment process is the predatory activity, exerted by them on bacteria that are suspended in the medium (LaRivie′re,1977). Hence, bacteria are dispersed (the ones that are not a fragment of floc) instead of normal removal in sedimentation. In terms of organic matter, suspended solids, and pathogens, they play a role in the deterioration of the effluent. Final effluent's quality is improved by the role that protozoa play on bacteria (O'connor &Dobbins, 1958; Horan, 1990).

Compared to stalked ciliates, greater requirements of food are there for free-swimming ciliates, as most of their energy is utilized in locomotion. Stalkedciliates show predominance when the population of free-swimming ciliates declines, at this point they can feed on the available bacteria in floc (Parkhurst&Pomeroy, 1972; Orhon&Artan, 1994).

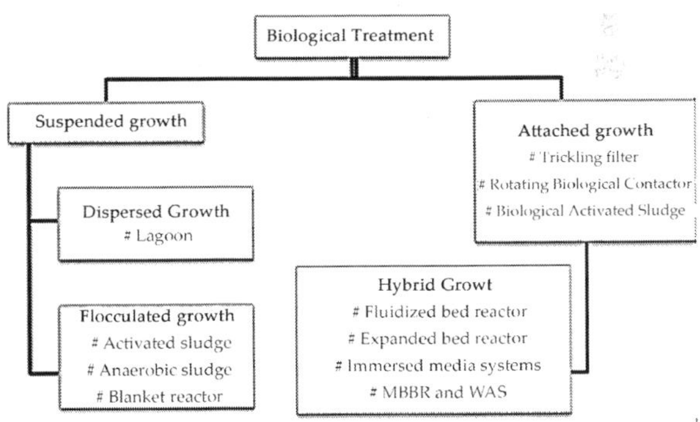

Figure 2.6: Biomass growth examples.

[Source: https://www.researchgate.net/figure/Figure-2-Biomass-growth-systems-in-wastewater-treatment-systems-Jianlong-et-al-2000_fig2_301363620]

2.5.3 Attached and suspended biomass growth

We can divide the treatment process of biological sewage (depending on biomass' structural formation) into the configurations listed below (in Figure 2.6). The list is ordered following the prevailing mechanism, though the mechanism of dispersed and attached growth can take place simultaneously (Po¨Pel,1979; Qasim, 1985).

Dispersed growth: the growth of biomass takes place in a liquid medium, in dispersed form and having no supporting structure

Systems:

i stabilization of variants and ponds

ii activated variants and sludge

iii The upward flow of sludge blanket reactors (anaerobic). They receive wastewater that contains suspended solids

Attached growth: a biofilm is formed as the growing biomass is attached to some support medium. This supporting medium can receive intermittent or continuous liquid discharges or can immerse in the medium (liquid). This support medium can be anything in solid naturals such as soil, sand, stones or it may be consisting agglomerate of biomass in the shape of granules or it can be some artificial plastic-type material.

Systems consisting of solid support for attachment:

i Trickling filters anaerobic filters

ii rotating biological contactors

iii land disposal systems

iv trickling filters

v submerged aerated biofilters

Support systems for attachment comprising agglomerated biomass.

i upward flow of sludge blanket reactors anaerobic

For biomass support systems, biological treatment principles are the same; inter convenience of specific aspects influence the treatment kinetics. Theoretical development is greater when we take it in terms of modeling, about dispersed growth with aerobic treatment. In years, a lot of researches on the activated sludge process is going on and also on the model formulation for dispersed-growth.

2.5.4 The Role of Biological floc in the System of Dispersed-growth

In activated sludge-like, few treatment processes, the organisms concentrate forming a structural unit that is broader and called a floc. Though in BOD removal microorganisms serve as agents in activated sludge, flocs play a major role in the process of removing organic matter.

Organic matter is stabilized by heterotrophic organisms, this result in increasing the efficiency of the activated sludge process. Another important property of microorganisms is that they organize themselves in form of structural units, this unit of floc can be separated from a liquid medium, and this can be done by sedimentation through simple physical means.

This separation serves responsibility in clearing the final effluent (with lessen concentrations of organic matter that is suspended). Final effluent quality is characterized by the low value of suspended BOD and soluble BOD. The mechanism of flocculation is reached in water treatment at the cost of adding chemical products, which take place in the treatment of biological sewage entirely through natural mechanisms (Grady&Lim, 1980; EPA, 1993).

A heterogeneous structure is represented by flocs, this structure comprises produced microbial material, inert material coming from sewage, organic matter absorbed, and dead and alive cells. The floc size is managed through the balance between the force of shear-stress (which is an effect of agitation and artificial aeration) and the forces of cohesion (LaRivie´re, 1977). Among all the microorganisms that are involved in the constitution of floc, (excluding protozoa and bacteria) rotifers, sometimes insect larvae, nematodes, and fungi can be found (Branco,1978).

Almost 40% of the particulate and soluble BODs that enter in the biological reactor through the means of ionic interactions can be absorbed by the flocs. Exoenzymes hydrolyze the particulate material before its absorption and metabolization, by the bacteria.

Because floc size varies between 50 to 500μm, a BOD marked gradient will be there along with a concentration of oxygen from the floc's external border where values are large, towards the center where the value of BOD is low. Bacteria may be deprived of nutrient resources while moving towards the center of floc, and this may reduce the viability of them (Handley, 1974; Horan,1990).

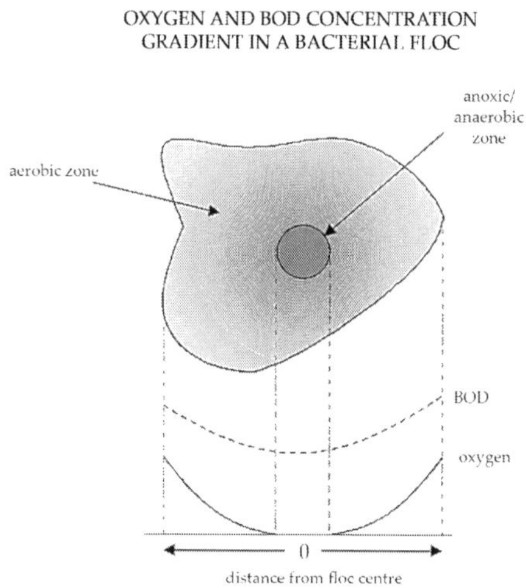

Figure 2.7: Oxygen and BOD gradients besides a typical *floc* (adapted from Horan, 1990).

[Source: https://www.iwapublishing.com/sites/default/files/ebooks/9781780402093.pdf]

The possible absence of nutrients and oxygen inside the floc should be considered, when their availability is analyzed in the liquid medium. This lies in support of the fact that in some cases, assumption of anoxic conditions is made, though, in a liquid medium, a minor concentration of DO (0.5 mg/L) may be found. A large portion of floc is deprived of oxygen, though the liquid medium is not. In this circumstance, in the floc's interior, the anoxic conditions are prevailing and in the oxygen absence, biochemical reactions occur. The oxygen gradients and the BOD along the floc are illustrated in Figure 2.7 (Huisman,1978; Hanisch, 1980).

The reasons due to which microbial growth is in floc form instead of in the form of cells that may be suspended freely in the medium are not well known. A prominent hypothesis for the floc structure is that the function of the structural matrix is exerted by the filamentous bacteria, it adheres to the floc-forming bacteria. It is generally accepted that through exopolysaccharides (in form of gelatinous layer or capsule), the attachment occurs. It was the attribute of *Zoogloea ramigera* only in the past, but indications are there that the gelatinous layer is produced through various genera that may include *Pseudomonas*. Exopolymers' continuous production, adheres to colloidal particles and other microorganisms, as a result, the size of floc increases. The floc is colonized by the adherence of protozoa, some evidence support that viscous mucus is excreted by them and this mucus plays a part in the cohesion of the floc (Horan,1990). The schematic of activated sludge floc's structure is shown in Figure 2.8 (Iawprc, 1987; Iwa,2000).

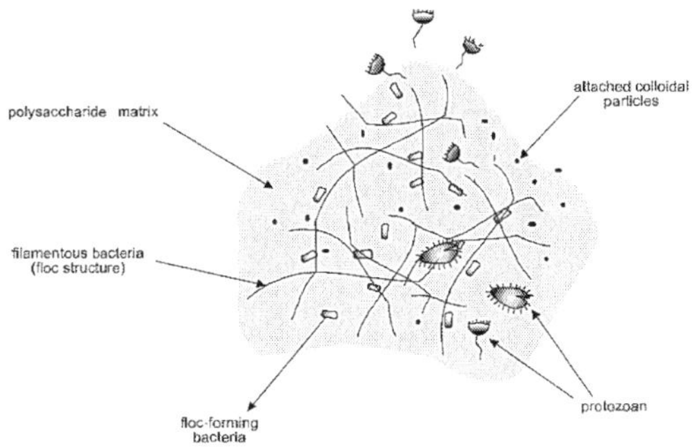

Figure 2.8: An activated sludge *floc's* typical structure (adapted from Horan, 1990).

[Source:-https://www.iwapublishing.com/sites/default/files/ebooks/9781780402093.pdf]

There is a delicate balance between floc-forming and filamentous organisms; and activated sludge plant's operational success depends on it. The basic three conditions that can take place are (Horan,1990):

1. *Equilibrium between* floc-*forming and filamentous organisms.* Good thickening and better settling ability of the sludge (Iwai &Kitao, 1994).
2. *Floc-forming organism's predominance.* In the floc, there is insufficient rigidity, which generates a floc that is weak and small and has poor settleability. Its consequences in so-called pin-pointfloc.
3. *Filamentousorganism's predominance.* Impeding the other floc's adherence, outside the floc, the filaments extend themselves. Hence, after the sedimentation process, a large volume is occupied by the flocs, this large volume can result in operational problems of the secondary sedimentation tank. As a consequence, the final effluent's quality is deteriorated. Such a condition is termed as sludge bulking (Keinath et al., 1977; Keinath, 1981).

2.5.5 Biofilm in the System of Attached Growth

The microorganisms' attachment to a suspended or solid supporting medium is termed as the process of immobilization. It has the benefit that it enables a high concentration of biomass to be retained for a long time, in the reactor (KoNig, 1990). Practically, almost all microorganisms are capable to adhere to a medium (supporting one) by producing extracellular polymers; they permit physical-chemical attachment. Recently, the technological application of processes of cellular absorption has started to be employed at an optimized and wider scale in numerous sewage treatment; and biotechnological processes (Lubberding,1995).

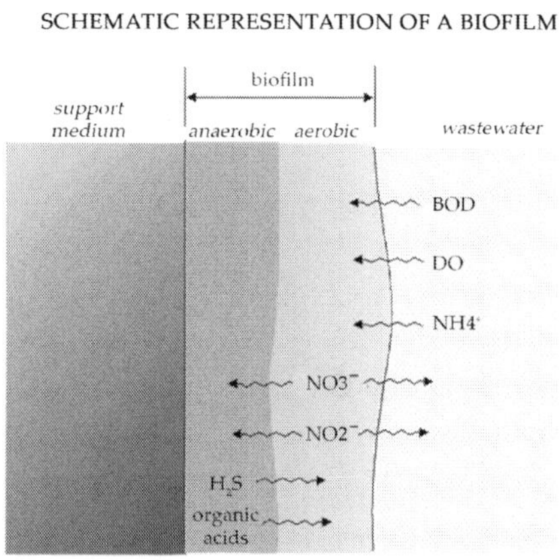

Figure 2.9: A biofilm's schematic representation (adapted from Iwai and Kitao, 1994).

[Source: https://www.iwapublishing.com/sites/default/files/eb-ooks/9781780402093.pdf]

Cell-to-cell interaction influences the attachment; because the polymer molecules are present at the surface and also through the medium concentration.(Rouxhet and Mozes,1990).

Onto the surface, the compounds in the biofilm that are elementary for the development of bacteria, such as oxygen, micronutrients, and organic matter are absorbed. After adherence, they are transferred by the biofilm, by the mechanism of diffusion; and here microorganisms metabolize them. Through the biofilm, diffusion of suspended or colloidal solids cannot take place; and are required to be hydrolyzed into small molecules. The metabolic products are transferred in the inversing direction, towards the liquid phase

(Sawyer&McCarty,1978; Iwai and Kitao, 1994). The operating mechanism of biofilm in the treatment process of sewage is illustrated in Figure 2.9.

Until anaerobic or anoxic conditions are reached, oxygen is consumed in the anaerobic reactor as it penetrates the biofilm. Hence, the internal layer without the presence of oxygen and external layer with the presence of oxygen might be found. In establishing layers, DO serves as a determining factor. In the anoxic layer, the reduction of nitrate will occur. In an anaerobic environment, organic acid formation and sulfate reduction will take place. This coexistence among anoxic, anaerobic, and aerobic conditions is a fundamental attribute of biofilm systems (Silva &Mara, 1979; Iwai and Kitao, 1994).

At three stages, the process of biofilm formation can be understood (IwaiandKitao,1994). The main attributes of these stages linked with the biofilm thickness are presented in Figure 2.10 and Table 2.8 (Stephenson, 1985).

Table 2.8: Biofilm formation stages (Iwai and Kitao, 1994)

Biofilmthickness	Attributes
Thin	• Supportmedium surface is not covered and the film is always thin • A logarithmic rate is followed by bacterial growth • Underthesameconditions, all microorganisms grow. To an extent, this growth is similar to the growth of dispersed biomass
Intermediate	• Greater film thickness • The rate of bacterial growth becomes constant • Theactivelayerthickness remains unaltered, not relying on the increase observed in thickness of thebiofilm • In limited organic matter supply, a metabolism is adopted by microorganisms that do not support growth but are sufficientfor maintenance • the thickness of film decreases when the maintenance requirements are higher than organic matter supply
High	• the biofilmthickness approaches to a very high level • The decay of the organisms counteract the microbial growth, by shearing stress and by up taking other organisms • From the support medium, the biofilm parts can be dislodged • Clogging occurs if growing biofilm does not dislodge from support medium

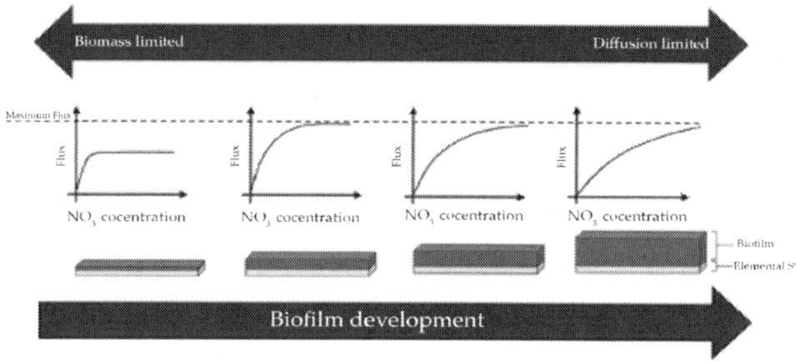

Figure 2.10: Substrate's concentration gradients (S) in biofilms differing in thicknesses (Lubberding, 1995).

[Source: https://onlinelibrary.wiley.com/doi/abs/10.1002/bit.27094In sewage treatment, when attached growth and dispersed growth are analyzed, comparing doubling time of cell and hydraulic detention time, is an important aspect.

For the growth of the microbial population, referring to a dispersed system of growth, the doubling time has to be less than the hydraulic detention time of the microorganisms. The doubling time is the time that is necessary for generating new cells and the hydraulic-detention time is the average time of water molecule staying in the system. If the cell doubling time is greater than the hydraulic detention time, then from the system, bacteria are *"washed"* out. It serves as a limiting factor in the sizing of biological reactors, provided that detention time and reactor volume are considered to be directly related.

In systems having biofilms, the cell doubling time can be greater than the hydraulic detention time, in such systems cells are not washed out due to the reason that bacteria are directly attached with a support medium. As an outcome, for reactors, lower volumes can be adopted.

Considering the comparison between attached-growth and dispersed-growth systems, the following are aspects relative to growth systems (IwaiandKitao, 1994; Lubberding,1995):

1. It is possible to operate the reactor with the cell doublingtime greater than hydraulicdetentiontime.
2. Compared to disperse- growth systems, *active* biomass concentration can be higher.
3. Compared to disperse- growth systems, the rate of substrate removal can be higher.

4. Compared to disperse- growth systems, the coexistence factor between anaerobic and aerobic microorganisms is greater due to the reason that the biofilm thickness is most of the times greater than biologicalfloc's diameter.
5. Substrates are found to be in the state of liquid but the cells are solid phased and are fixed. This reduces the need for subsequent clarificationstage.
6. The microorganisms are frequently reutilized while in the dispersed-growth system, the reuse can be implemented via recirculating biomass.
7. There are limitations if the thickness of biofilm is high and these limitations come in the act when the substrate diffuses into biofilm (Lubberding,1995).

76. Silva Jr., C., Sasson, S. Biology 2: Living Beings, Structure And Function. (Cesar And Sezar). Current Publisher, Sa˜o Paulo, 2,377-382.
77. Silva, S.A., &Mara, D.D. (1979). Biological Treatments Of Wastewater: Stabilization Ponds. Abes, Rio De Janeiro, 1, 130-140.
78. Sobsey, M. D. (1989). Inactivation of health-related microorganisms in water by disinfection processes. Water science and technology, 21(3), 179-195.
79. Sohn, L. L., Saleh, O. A., Facer, G. R., Beavis, A. J., Allan, R. S., & Notterman, D. A. (2000). Capacitance cytometry: Measuring biological cells one by one. Proceedings of the National Academy of Sciences, 97(20), 10687-10690.
80. Stephenson, R. V. Et Al (1985). Performance Of Surface Rotors In An Oxidation Ditch.Journal Of Environ. Eng. Div., Asce, V.111, N.1, P 79–91.
81. Tchobanoglous, G. & Schroeder, E.D. (1985). Water Quality: Characteristics, Modeling, Modification (Vol.1, Pp. 1-32). Addison-Wesley, Reading, Ma.
82. Thurnheer, T., Köuhler, T. H. I. L. O., Cook, A. M., & Leisinger, T. (1986). Orthanilic acid and analogues as carbon sources for bacteria: growth physiology and enzymic desulphonation. Microbiology, 132(5), 1215-1220.
83. Tsuge, T. (2002). Metabolic improvements and use of inexpensive carbon sources in microbial production of polyhydroxyalkanoates. Journal of bioscience and bioengineering, 94(6), 579-584.
84. Van Haandel, A.C., &Lettinga, G. (1994). Anaerobic Sewage Treatment. A Manual For Hot Climate Regions, 1, 1-43.
85. Viesmann Jr, W., &Hammer, M.J. (1985). Water Supply And Pollution Control. Harper And Row,Publ. New York, Vol. 4, Pp. 790-797.
86. Vinuselvi, P., Kim, M. K., Lee, S. K., & Ghim, C. M. (2012). Rewiring carbon catabolite repression for microbial cell factory. BMB reports, 45(2), 59-70.
87. VonSperling,M.(1983a).The Influence Of The Delay Time In Determining The Coefficient Of Deoxygenation. Sanitary Engineering,22 (3),Jul1983.P.375–379.
88. Von Sperling, M. (1983b). Self-Cleaning Of Water Courses, Masters Dissertation, Des-Ee.Ufmg, Belo Horizonte, 1, 350-366.

89. Von Sperling, M. (1990). Optimal Management Of The Oxidation Ditch Process. Phd Thesis, Imperial College, University Of London, 1990, 1, 360-371.
90. VonSperling,M.(1993).Determination Of The Oxygenation Capacity Of Activated Sludge SystemsIn Operation. In: Brazilian Congress Of Sanitary And Environmental Engineering, 17, Natal, 19–23 Setembro 1993, Vol. 2, Tomo I, pp. 152–167 (In Portuguese).
91. Von Sperling, M. (1994). A New Unified Solids flux-Based Approach For The Design Offinal Clarifiers. Description And Comparison With Traditional Criteria. Water Science Andtechnology, 30 (4), pp 57–66.
92. Von Sperling, M. (1997). Principles Of Biological Treatment Of Wastewater. Vol. 4. Activated Sludge. Department Of Sanitary And Environmental Engineering, Ufmg, 1, 420-428.
93. Von Sperling, M. (1999). Performance Evaluation And Mathematical Modelling OfCo- Liform Die-Off In Tropical And Subtropical Waste Stabilization Ponds. Water Research, **33** (6). Pp.1435–1448.
94. VonSperling,M.(2002).RelationshipBetweenfirst-OrderDecayCoefficientsInPonds, According To Plug flow, Cstr And Dispersed flow Regimens. Water Science And Tech- Nology, **45** (1). Pp.17–24.
95. VonSperling,M.,&Fro´Es,C.M.V.(1999). DeterminationOfTheRequiredSurfaceArea ForActivatedSludgefinal ClarifiersBasedOnAUnifiedDatabase.WaterResearch,**33**(8). Pp.1884–1894.
96. Wagner, M., Loy, A., Nogueira, R., Purkhold, U., Lee, N., & Daims, H. (2002). Microbial community composition and function in wastewater treatment plants. Antonie Van Leeuwenhoek, 81(1-4), 665-680.
97. Wanner,J.(1994).ActivatedSludgeBulkingAndFoamingControl. TechnomicPublishing Company, 1, 320-327.
98. Wef & Asce (1992). Design Of Municipal Wastewater Treatment Plants. Water Environ- Ment Federation / American Society Of Civil Engineers, 1, 1590-1592.
99. Werker, A. G., Becker, J., & Huitema, C. (2003). Assessment of activated sludge microbial community analysis in full-scale biological wastewater treatment plants using patterns of fatty acid isopropyl esters (FAPEs). Water research, 37(9), 2162-2172.
100. White, M.J.D. (1976). Design And Control Of Secondary Settlement

Tanks. Water Pollution Control, **75**, Pp 459–467.

101. Wilson, F. (1981). Design Calculations In Wastewater Treatment. E.&F.N.Spon, London, 1, 210-221.
102. Wpcf & Asce (1988). Aeration. A Wastewater Treatment Process, 1, 160-167.
103. Yi, C., Li, C. W., Ji, S., & Yang, M. (2006). Microfluidics technology for manipulation and analysis of biological cells. Analytica Chimica Acta, 560(1-2), 1-23.
104. Zhao, X., Wang, Y., Ye, Z., Borthwick, A. G., & Ni, J. (2006). Oil field wastewater treatment in the biological aerated filter by immobilized microorganisms. Process Biochemistry, 41(7), 1475-1483.

CHAPTER
3

AEROBIC WASTEWATER TREATMENT AND UNITS

CONTENTS

3.1 INTRODUCTION

Microorganisms that crop up naturally are the backbone of the treatment of wastewater. A lot of intricate compounds incorporated with household wastewater are thrived by the organisms involving fungi, bacteria, rotifers, protozoa, and additionally microbes. The very well-engineered bioreactors are being the secondary-treatment methods that are discovered at municipal wastewater treatment plants (Lee & Welander, 1996). The microbes are supplied by these bioreactors with the best possible circumstances to facilitate the restoration of domestic wastewater. Rapid oxidization of bioavailable organic, soluble, and nitrogenous compounds can be done with the mechanical inclusion of aerobic, facultative microbes, and dissolved oxygen (La Motta et al., 2007; Ichinari et al., 2008).

In general, the aim of this module is: (i) to provide details of numerous engineered systems that uphold high-rate digestion, (ii)to provide an analysis of the biochemical oxidation of colloidal organic compounds and soluble with the help of aerobic microbial digestion, and (iii) to provide an insight of the service and maintenance needed to carry these systems operational (U.S. EPA, 2000; Najafpour et al., 2006).

This module is comprising of 2 sections: (1) the aerobic treatment procedure and (2) aerobic treatment units. The summary of the biochemical oxidation of soluble organic compounds is discussed in section one which is settled in domestic wastewater (Perle et al., 1995; Jianmin & Fangwei, 2005). Two categories of the scheme of biological treatment units can be suspended-growth and attached-growth. In turning the organic carbon into inorganic carbon, the bio-processes are identical in both classes.Citations given in the module provide directions to the readers to the textbooks from where more precise clarifications can be reviewed about processes regarding biological wastewater treatment (Wang et al., 2006; Thanh et al., 2008).

For decentralized wastewater management, the use of aerobic treatment systems is described in the following part of this chapter.Although all wastewater treatment equipment that is designed to sustain aerobic conditions are treated as aerobic treatment units, two classifications are suggested by the professionals of the community of onsite wastewater management i.e. non-saturated and saturated (with wastewater) (Suzuki et al., 1978; Heddam et al., 2016).Any unit that holds aerobic and saturated conditions regardless of attached-growth or suspended-growth is normally known as an "ATU" which is abbreviated as Aerobic Treatment Unit. The key focus of this section is the exploitation of ATUs as a high-rate, engineered wastewater treatment process. The atmospheric oxygen is indifferently transformed into solution in non-saturated attached-growth systems, because of the movement of water over or through the media. The remarkable packed-bed filters and trickling filters (i.e. those encountered at small-scale municipal wastewater treatment plants) represents this type of biological process (Fan et al., 1996; Schwarzenbeck et al., 2004).

3.2 BIOCHEMICAL WASTEWATER TREATMENT

The bacteria and other microorganisms are considered to be the unwanted elements of wastewater by most of the people.Indeed, true pathogenic are the small ratio of the microbes spotted in wastewater (Dogan et al., 2008; Abdalla & Hammam, 2014).The rise of naturally developed aerobic microorganisms is encouraged by the aerobic wastewater treatment as a tactic of renovating wastewater.The microbes like these are the weapon of wastewater treatment plants.The high energy modes of carbon are organic compounds (Comeau et al., 1986; Aivasidis & Diamantis, 2005).The ammunition that gives force to these engines is the carbon dioxide which is the oxidation of organic compounds to the low-energy form.One of the basic tasks of wastewater engineers is to figure out that how to mingle soluble organic compounds, dissolved oxygen, and aerobic microorganisms for high-rate oxidation of organic carbon (Dharmsthiti & Kuhasuntisuk, 1998; Mohan et al., 2008).

3.3 NATURAL PROCESS

The "decomposers" are the microorganisms that are capable of the complex organic compound's oxidation. The simple forms of carbon are returned to the water, atmosphere, and soil. These decomposers grow on the availability of high concentrations of organic pollutants. A very low organic strength is essential for the wastewater being released back into surface water bodies as the microorganisms residing in natural water bodies are the same (Wang & Peng, 2010; Ouyang et al., 2015).A sufficient accumulation of dissolved oxygen is necessary for natural aquatic systems to reinforce advanced life kinds i.e., macroinvertebrates, fish, etc. The aerobic conditions are preferred by most of the decomposing microbes instead of anaerobic conditions. The aerobic decomposition of organic compounds uses up dissolved oxygen drained from the water on the accessibility of dissolved oxygen. The concentration of dissolved oxygen will decline down the level required to preserve a feasible aquatic system if the amount of re-aeration and the amount of consumption are not equal (Zhang et al., 2013; Freitas et al., 2015).

Generally, the soluble concentration, bioavailable organic compounds found in water is determined either as cBOD or carbonaceous biochemical oxygen demand. It is already explained that the aerobic microorganisms employing dissolved oxygen results in the demand for oxygen because of the decomposition of nitrogenous compounds and organic carbon. The oxygen is provided to the aerobic microorganisms in the engineered wastewater biochemical oxidation, thus the substrate (organic carbon) will be consumed to give power to their metabolism. The outcome is the change of organic pollutants into new microbial cells and inorganic compounds. An aggregation of biological substances will be formed by the net production of cells i.e. production of new cells versus the destruction of old cells (Wang et al., 2011; Anagnostopoulos & Vavatsikos, 2012).

Conventional organic stuff that is turned up at residential intensity wastewater consists of fats, carbohydrates, proteins, soaps, detergents, and urea. The hydrogen, oxygen, and carbon are contained in all of these compounds. Organically bound nitrogen, phosphorus, and sulfur are also incorporated with domestic wastewater (Amuda & Ibrahim, 2006; Kumar et al., 2015).These three mentioned components are biologically converted from organic to mineralized forms during degradation of biochemicals (i.e., NH4, NH3, NO3, PO4, andSO4).

3.4 MICROBIAL METABOLISM

The aggregation of the biochemical processes that are engaged in the strengthening the cell protoplasm (anabolism), and in the devastation of organic compounds (catabolism) is metabolism.The conversion of chemically-bound energy into energy forms is done by these processes that are applicable in the processes of life-sustaining (Cerniglia,1984; Tang et al., 2013).The degradation process of exothermic, oxidative, enzymatic that leads to the discharge of free energy from the architecture of wide organic molecules is known as Catabolism. For the creation of new organic material, a few of the discharged energy is available.A synthesis process derived from the expansion of size and complication of the structure of organic chemicals is anabolism (Benefield & Randall, 1985; Chubukov et al., 2014).

3.4.1 Fermentation and Respiration

To lessen the complex organic compounds to simple organic forms, the fermentation process is used by anaerobic heterotrophic and aerobic microorganisms. The organic carbon is used by heterotrophs organisms for the composition of new biomass. These microorganisms rely upon a source of organic carbon that is quickly available for chemical energy and cellular synthesis because they are decomposers and consumers (Taylor, 1942; Verduyn et al., 1992).In wastewater treatment, they are the basic captives in the soluble body oxidation. In contrast, a cellular substance can be created by autotrophic microorganisms from simple forms of carbon i.e. carbon dioxide. These organisms lie at the ground of the food chain. They are not dependent on other organisms for the formulation of intricate organic compounds. For the elimination of nitrogen from wastewater, autotrophic microorganisms are valuable (Stolz et al., 2006; Schuetz et al., 2012).

Fermentation is the enzymatic, exothermic disruption of soluble organic compounds as demonstrated in equation 1, and is not dependent on the availability of dissolved oxygen. Normally, two stages describe the fermentation process i.e. methane fermentation and acid fermentation. The outcome of the acid fermentation process involves alcohol and volatile fatty acids (VFA). As yet much of the carbon is in an organic form, the slight contraction in BOD appears (Egami, 1973; Kerner, 1993). While doing methane fermentation, a part of the acid-fermentation outputs is transformed into carbon dioxide gas and methane. The reduction in BOD is the result of this transformation. Anaerobic organisms are restricted to the fermentation process. Therefore, the production of methane can only be possible with anaerobic conditions (Blair et al., 1985; Passoth et al., 1996).

Equation 1

$$\underset{\text{organic compounds}}{\text{COHNS}} \xrightarrow[\text{microbes}]{\text{heterotrophic}} \begin{bmatrix}\text{volatile}\\ \text{fatty}\\ \text{acids}\end{bmatrix} + CO_2 + H_2O + CH_4 + \text{energy} + \text{residuals}$$

The aerobic microorganisms can additionally convert the VFA (and other bioavailable organic compounds) into water, carbon dioxide, and additional energy by the respiration process (Lehninger, 1973). The existence of oxygen is required by respiration as shown in equation 2. For the catabolic deterioration of the VFA, oxygen behaves as an electron acceptor. The high-rate wastewater treatment can be provided by aerobic systems because aerobic microbes can quickly turn bioavailable organic carbon into inorganic carbon (Zhang et al., 2017).

Equation 2

$$\begin{bmatrix}\text{volatile}\\ \text{fatty}\\ \text{acids}\end{bmatrix} + O_2 \xrightarrow[\text{microbes}]{\text{aerobic}} \text{energy} + CO_2 + H_2O + \text{residuals}$$

3.4.2 Biosynthesis

The most crucial and complicated energy-demanding exercise of all living organisms is biosynthesis following the Lehninger (1973). It is exhibited in equation 3 that biosynthesis is the process of forming distinguishing chemical components of cells from smooth harbingers, and the compilation of these components into structures e.g. contractile elements, the membrane systems, mitochondria, ribosomes, and nuclei(Kende, 1993; Vogt, 2010). There are two kinds of elements needed for the biosynthesis of cell components: (1) to set up the harbinger into covalently-bonded cellular structure, adenosine triphosphate (ATP) and other kinds of chemical energy are required and (2) precursors that keep the hydrogen, carbon, nitrogen, carbon, and other components discovered in cellular structures (Delmer, 1987; Aura,2008).

Equation 3

$$\underset{\text{precursors}}{\text{simple}} \xrightarrow[\text{energy}]{\text{microbes}} \underset{\text{new cells}}{C_{60}H_{87}N_{12}O_{23}P}$$

The C60H87N12O23P given in equation 3 represents cell composition. The fundamental composition of a cell is usually written as C5H7NO2 while not considering phosphorus (Chipasa, 2003; Sui et al., 2011). It is influential to strengthen the point that from the wastewater stream, the cellular components are being captured and so, new cells are formed from much of the wastewater elements conversion. The common bacterial cell composition is listed in Table 3.1 (Delmer & Amor, 1995; Von Wettstein et al., 1995).

Table 3.1: Percent elemental composition of cellular material (Metcalf and Eddy, 2003)

Carbon 50.0	Oxygen 22.0	Nitrogen 12.0	Hydrogen 9.0	Phosphorus 2.0	Sulfur 1.0
Potassium 1.0	Sodium 1.0	Calcium 0.5	Magnesium 0.5	Chlorine 0.5	Iron 0.2
Other trace elements including Zn, Mn,Mo, Se, Co, Cu, and Ni: 0.3					

3.4.3 Endogenous Respiration

Rather than the production of new cells, the microbes will nourish each other at high-rates under substrate-restricted circumstances. It is exposed in equation 4 that the endogenous respiration is being the aerobic declination of cellular material. It is not 100 percent effective and, in this manner, there is an agglomeration of lethargically-degradable cellular substance and other residues (Reynolds, 1982; Pomiès et al., 2013). It will be discussed in thereafter that in the endogenous respiration stage, the aerobic treatment units working in the distributed wastewater management corporation is operated (Schaar et al., 2010; Westerhoff et al., 2013). The "extended aeration" is referred to the process which supplies much enough aeration to make sure that the microbes would start fueling each other, once the food is depleted. This response reduced the size of accumulated biomass that the maintenance provider must evacuate (Wei et al., 2003; Joss et al., 2005).

Equation 4

$$\underset{\text{cellular material}}{C_{60}H_{87}N_{12}O_{23}P} + O_2 \xrightarrow[\text{microbes}]{\text{aerobic}} CO_2 + H_2O + PO_4 + NH_3 + \text{residuals}$$

3.5 ENVIRONMENTAL EFFECTS

An environment must be arranged for microorganisms that permit them to flourish to present high-rate oxidized organic pollutants. The natural selection, durability, and production of microorganisms are overwhelmed bypH, temperature, dissolved oxygen, and other aspects along with their rate of biochemical oxidation (Ivery, 1996; Roeder & Brookman, 2006).

3.5.1 Temperature

The role of temperature is the rate of bio-oxidation. For cell synthesis and survival, numerous microbial species have optimum temperatures.

1. In a temperature range of -2° to 30°C (28° to 86°F), psychrophilic microorganisms are prospered. While 12° to 18°C (54° to 64°F) is the optimal temperature.
2. In a temperature range of 20° to 45°C (68° to 113°F), mesophilic microorganisms are grown. Where 25° to 40°C (77° to 104°F) is the optimum temperature.
3. In a temperature range of 45° to 75°C (113° to 167°F), thermophilic microorganisms have thrived. Whereas 55° to 65°C (131° to 149°F) is the optimum temperature.

In general, microbial activity increases with the increment in temperature. All over, distributed aerobic treatment units are concealed and for the heat generation, the soil serves as a sink by the exothermic activity inside the treatment unit. a mixture of Mesophilic and Psychrophilic organisms will be contained by the microbial population in a submerged aerobic treatment device (Larson & Safferman, 2009; Heger et al., 2010).

3.5.2 Food to Microorganism Ratio (F/M)

This ratio indicates the loaded mass of bio-available organic compounds (substrate) into the aeration chamber every day in contrast to the mass of microorganisms already present within the aeration chamber. Commonly, this ratio is indicated in terms of the mass of BOD per day per mass of microbes contained in the treatment unit (Crites and Tchobanoglous, 1998). There is a dynamic microbial population and it retorted to variations in life-sustaining parameters. In organic loading, abrupt changes are responded by the lapse of time to revisions in the microbial population. After all, the increase in organic loading is responded by a briskly increase in the population if the remaining factors are constant. The basin's hydraulic retention time should correlate to the time needed for the population to grow to productively handle the increased organic load. Still, the increase in organic loading is usually related to the increase in hydraulic loading. The constant residence time will not have been taken by the effluent, neither they have been disclosed to the uniform concentration of microbes if and only if the ways of flow equalization has not been specified.

3.5.3 Acid Concentration

The wastewater treatment has been a notable impact by the influent pH. It is reported by Benefield and Randall (1985) that there is a possibility of organic wastewaters treatment over a broad pH range, but still, for microbial growth, the optimum pH is at intervals of 6.5 and 7.5. It is amusing to notice that bacteria utmost raised at lightly alkaline water. Likewise, fungi and algae outstandingly produced in slightly acidic water. The changes in enzymatic activity are the leading reason for the response to pH.

3.6 AEROBIC TREATMENT UNITS (ATUS)

The high-rate oxidizers of nitrogenous compounds and soluble organic are referred to as aerobic treatment units. Any new processes are not exploited by ATUs in a biological point of view, that do not utilize earlier in broad-scale wastewater treatment plants (Gavrilescu & Macoveanu, 1999; Bélineet al., 2007).ATUs have unique technology for narrow flow cases is the packaging and design of these systems. These tools are scaled-down wastewater treatment plants. As aerobic digestion declines BOD and nitrification converts ammonia, various commercially presents ATUs have further chambers that upgrade the removal of suspended solids, pathogens, and nutrients from the effluent (Van Lier & Boncz, 2002).Other exclusive aspects of the scheme of ATUs are that they are easy to install at remote locations and they are easy to maintain for semi-skilled maintenance providers.ATUs must be maintenance-friendly and dependable for those deployed at small commercial sites and homesites (Makinen t al., 1993; Pophali et al., 2007).

3.6.1 Process Description

Wastewater that is treated essentially gets into the aeration unit and dissolved oxygen is mixed with it and delayed or/and attached microbes. The organic compounds are converted into new cells, residual matter, and energy by aerobic microbes. A part of the biological solids is removed from the effluent and is kept inside the ATU because the water crossed over the clarifier. The biological solids recline into the aeration chamber to being attended as a seed for the production of new microbial. In the base of the chamber, residuals and settled biomass will assemble that must be cleaned out with regular maintenance (Mohan et al., 2005; Tanner et al., 2012).

The clarification is an essential part of producing a high-quality effluent as the biomass requires oxygen.The effluent's soluble BOD is normally lower to 5 mg/L, although the biomass solids carryover may provide an effluent BOD of 20 mg/L or more (Benefield and Randall, 1985).To endorse biomass separation, a conical-shaped clarifier is associated with many ATUs.The velocity of the fluid lessens with the increase in the cross-sectional area of up flow.The movement of biomass will not be towards upward if its settling velocity is higher than the fluid velocity.The biomass will lie down into the aeration chamber at the time of no flow.To extract the biomass from the effluent, other ATUs may integrate inline filters.These filters compel time to time maintenance to get rid of the growth of solids (Gouveia et al., 2015; Alvarino et al., 2016).

3.6.1.1 Typical ATU Configurations

Many ATUs functions as a complete mix tank, an intermitted-flow, constant volume reactors. As the influent is not constant, the flow is interrupted versus continuous. The fillings of the aeration chamber are completely blended to increase the association between microbes, wastewater, and dissolved oxygen. Effluent transfers from the aeration chamber to a clarifier. The inflow rate is directly responded to by the discharge rate. Sequencing the batch reactors is the exception to this induction. This treatment device works in batch mode which is discussed later in this section (Chan et al., 2009; Kassab et al., 2010).

3.6.1.2 Extended Aeration

A lot of commercially applicable ATUs function as extended aeration units. Extended aeration is symbolized by long detention times, long-term aeration, less biomass gathering, and less food to microorganism ratio.It is shown in figure 3.1 that the microbes would be in the self-growing phase and the bioavailable organic carbon will be consumed quickly, including biomass by giving good enough dissolved oxygen and minimum soluble organic matter (Stewart et al., 1962; Bertanza, 1997). The target is to stabilize the mass of cells internally biochemically-degraded per day and the mass of new cells incorporated per day. It is suggested by ASCE (1977) that 2000 cubic feet of air must be inserted in the water per pound of BOD5 removed for a treatment unit to work in extended aeration (Middlebrooks & Garland, 1968; Lo et al., 1994).

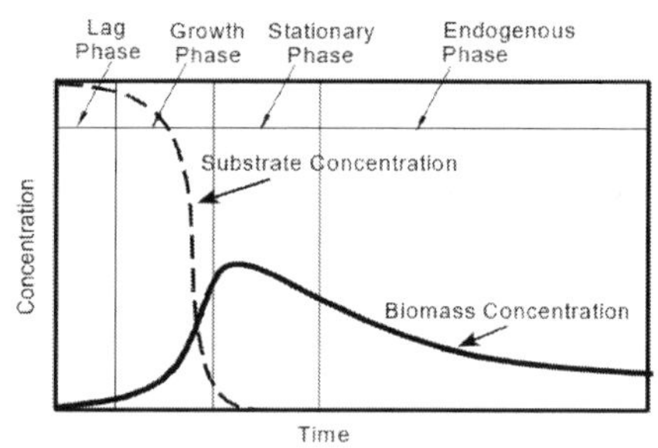

Figure 3.1: Kinetics of aerobic digestion: substrate increases with the increase in biomass.

Note:- *A "batch-style" function of the substrate is represented by these curves where biomass concentration variates with changes in substrate concentration. As the concentration of substrate turns to be nearly constant, therefore intermitted flow, complete mix systems only perform over a small range on these curves.*

[Source: http://onsite.tennessee.edu/Aerobic%20Treatment%20&%20ATUs.pdf]

3.6.1.3 Suspended Growth Bioreactors

Figure 3.2 demonstrates that suspended growth aerobic treatment units are scaled-down triggered sludge plants. Activated sludge is referred to as a heterogeneous microbial culture mostly constitutes of protozoa, bacteria, fungi, and rotifers (Sombatsompop et al., 2006). It is the bacteria pledged for adjusting many of the organic material, while the rotifers and protozoa (serving as killers) are essential in clear away the scattered bacteria that alternatively would leave in the ATU's effluent (Benefield and Randall, 1985). The biomass is utterly combined with the biodegradable organic compounds. Distinctive organisms are clumped collectively (flocculate) to design an active mass of microbes i.e. biological floc (Davis and Cornwell, 1991). This slurry of wastewater and biological floc is known as mixed liquor (Reynolds, 1982; Lee et al., 2001).In the mixed liquor, the concentration of microorganisms is determined as the mg/L of mixed liquor volatile suspended solids (MLVSS).

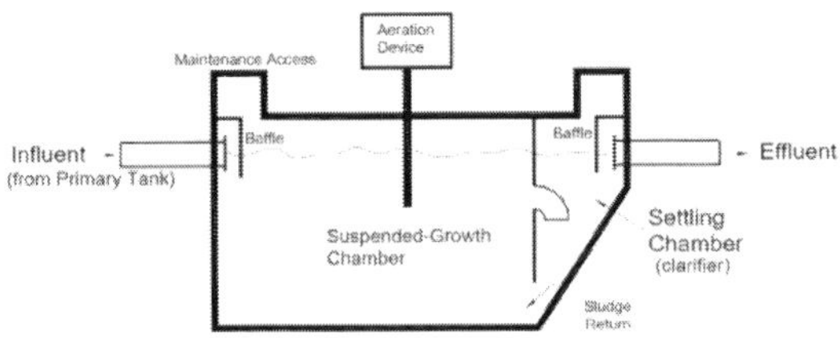

Figure 3.2: Diagram of the scheme of a suspended-growth ATU.

[Source: http://onsite.tennessee.edu/Aerobic%20Treatment%20&%20ATUs.pdf]

The term "activated" introduced by Reynolds (1982) is used to express the reactive character of the biological solids. The organic solids are adsorbed by the suspended floc and soluble organic compounds are absorbed by it on entering the wastewater into the aeration chamber. The organic solids are solubilized during enzymatic activity. Once in solution, the biochemical oxidation oxidizes soluble organics. The capability of the biological solids to absorb and adsorb substrate is speedily filled at the ATU's inflow. When the mixture approaches the clarification zone, then with the progress of the oxidation process the biological solids or "activated" sludge arebeing re-activated. The biological solids are substrate bounded at the ATU's downstream end, and hence are highly reactive to the surplus dissolved and suspended organic solids (Tziotzios et al., 2005; Muñoz et al., 2008).

3.6.1.4 Attached-Growth Bioreactors

Attached growth systems areare another vast class of aerobic treatment which is usually known as fixed-film reactors, an immobile medium is granted for microbial attachment. ATU scheme with overwhelmed attachment sites specified for fixed-film growth is exhibited in figure 3.3. (Lee et al., 2001; Sombatsompop et al., 2006).

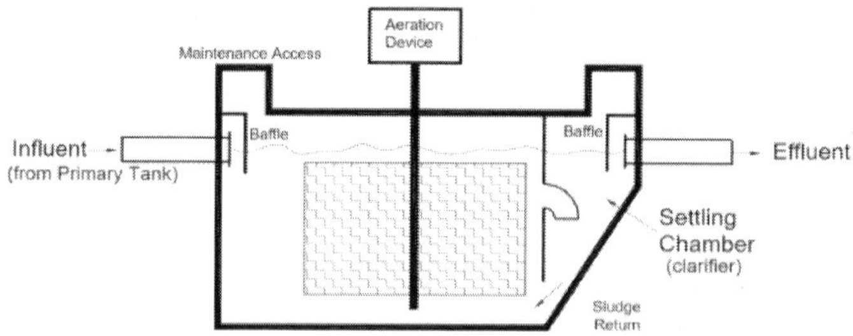

Figure 3.3: ATU Schemealong with submerged attachment sites supplies for fixed-film growth.

[Source: http://onsite.tennessee.edu/Aerobic%20Treatment%20&%20ATUs.pdf]

The biological film absorbs wastewater run out across or over the media, colloidal, fine suspended, and dissolved organic solids. The attached microorganisms are connected to the dissolved oxygen and wastewater by either traversing the media over the liquid or by liquid pumping spent the media. The presence of treatment units as suspended growth configured to integrate the attached-growth in the equal container. The consolidation of suspended growth and attached-growth processes strengthen the performance and scope of aeration units are mentioned as "coupled-contact aeration" (U.S. EPA, 2002; Achilli et al., 2011). A lower BOD and effluent suspended solids, and higher degree stability of microbial population is provided by this dual system strategy. For turbulent water, large channels are provided to move over the level and attached growth areas are immersed. Suspended-growth microbes are permitted by these large channels to prosper. Water circulation to the air-liquid interface or/and straightly injecting air provides aeration. Enormous attached-growth will descend and settle to the ground of the chamber. These solids will assemble and must be detached in the procedures of intermittent maintenance (Arya et al., 2016).

3.6.1.5 Rotating Biological Contactor (RBC)

The combination of attached-growth and suspended-growth bioprocesses can be possible by rotating biological contactors. A common shaft mounts at an order of closely spaced circular disks rotated slowly.

The shaft is situated either just below or merely above the water surface. The wastewater and air both expose the top of the disks allowed by this location during rotation. A typical disk is manufactured from an inert substance such as polyvinyl chloride or polystyrene.

The disks are attached to a fixed-film biological growth and the organisms are opened up to the food on submerging.

The air is carried the fixed-film by the reactor in rotation and absorbs oxygen. Because of the movement of the contactor surface back through the wastewater, the volume liquid is mixed with a large amount of dissolved oxygen (ASCE, 1977; Cema et al., 2007). Few of the excess biomass is trimmed off the disk with the increase in the density of the attached biomass on the disk. The rotation of the disks keeps this biomass in suspension (Metcalf and Eddy, 2003).

Eventually, the solids are carried out of the reactor chamber and inside the clarifier by the flow of wastewater. Commonly, an RBC submerges approximately 35 to 45% of the surface of the disks which is created with the shaft merely over the water surface. Almost 70-90% submergence is produced by a system that submerges the shaft in the water (Crites & Tchobanoglous, 1998; Pynaert et al., 2003).

3.6.1.6 Sequencing Batch Reactor Systems or Periodic Processes

The processes carried out in a sequence namely flow equalization, aeration, clarification, and biomass wasting in the same tank in the sequencing batch reactor (SBR) system (U.S. EPA, 1992). As the system needs to be closed as per the requirement of several SBRs to influent in the period of the treatment cycles.

Therefore, two reactors operated in parallel are recommended to retain continuous flow. After all, for continuous flow, single-tank reactors can be used with recent inlet designs (Irvine et al., 1989). The feature of flow equalization is provided by the SBR process which tends to balance the power and volume of wastewater inflow.

of energy utilized by the device. The combination of mixing with aeration is the most familiar way of increasing energy efficiency. Turbulent mixing is recommended to widen the convenience for microbes to get in contact with both dissolved oxygen and soluble organic compounds. The rate of transfer of oxygen is equal to the rate of microorganism's consumption for steady-state conditions. In the mixed liquor, dissolved oxygen must be retained at 1 to 3 mg/L. It is reported by Metcalf and Eddy (2003) that the dissolved oxygen of two to seven grams per day is required for each gram of MLVSS for residential power wastewater.

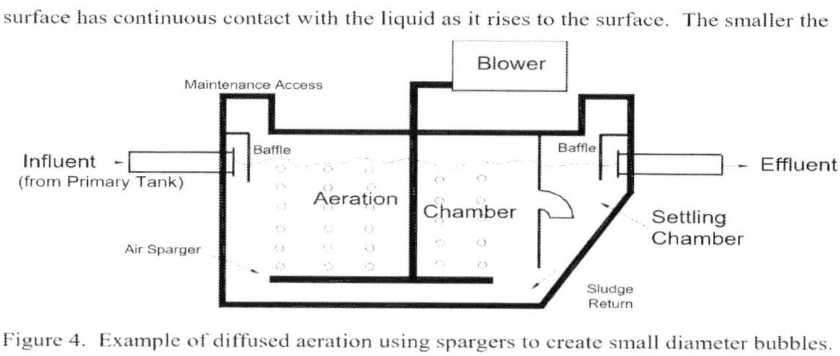

Figure 3.4: Example to create small diameter bubbles using diffused aeration using spargers.

[Source: http://onsite.tennessee.edu/Aerobic%20Treatment%20&%20ATUs.pdf]

Using porous ceramic diffusers is one of the methods of producing small bubbles. Many ends of outflow and a huge loss of air pressure are created by the small-interconnected paths within the ceramic matrix. Streams of small bubbles are produced by this combination over the covering of the ceramic diffuser.

To machine opening into plates and pipes is a second mechanism of injecting air. Jet aerators are used by numerous large-scale aerobic digesters. Airstreams help to transfer oxygen and to give energetic mixing of the basin constituents.

An aspirated mixer is the third kind of diffuse aerator. A mixing propeller is hooked up to a hollow shaft that is ventilated to the air demonstrated in figure 3.5. This paddle is situated close to the ground of the aeration chamber. On shaft spinning, a venturi-type effect produces a vacuum lower to the shaft and inserts air into the water.

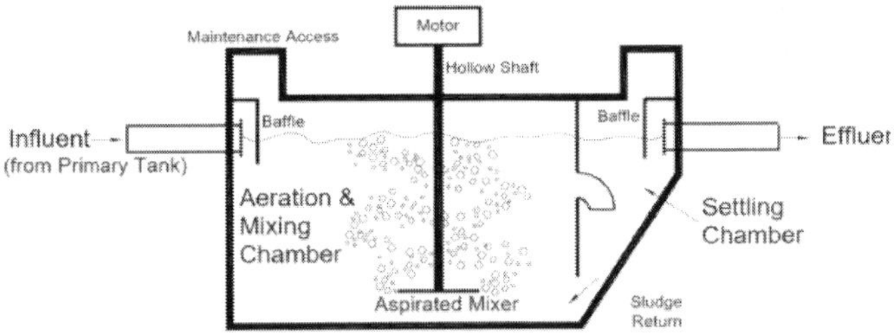

Figure 3.5: Aeration using a hollow drive shaft and an aspirated propeller.

[Source: http://onsite.tennessee.edu/Aerobic%20Treatment%20&%20ATUs.pdf]

While reducing the floc shearing, the balance of the need of agitation must be done by the mixing devices.Excess of shear may result in imperfect settling conditions in the clarifier.

Assorted ATU manufacturers exploit a cycled-aeration approach. Some energy savings are provided, and nitrogen evacuation (limited anoxic conditions) is promoted by cycling the aeration system.Because of non-flocculating microbes and gas flotation, poor settling biomass can be produced by this technique. Hence, precautions must be taken.

3.6.1.8 ATU Influent

The primary treatment must be given to the influent to the aerobic treatment chamber supported by either septic or another form of the primary tank. Before the entrance of influent into ATU, the separation can be provided by these tanks of easily floaTable 3. and setTable 3. solids.A huge portion of these solids is possibly either slowly degradable or non-degradable (Cao et al., 2017).

Guidance will be provided by the makers of ATUs as to the dimensions of a primary tank previous to their aerobic treatment device.A component of dilution is also provided by primary tanks to lessen the effects in the ATU of a chemical jolt to the microbial population.Before the entrance to the aerobic chamber, bleaches and additional disinfectants have to be supremely diluted.Medications are extremely poisonous to the microbial population e.g. chemotherapy drugs and antibiotics. A directory of products that should be avoided for the wastewater stream is prepared by most of the manufacturers (Larson& Safferman, 2009).

3.6.2 Operational Issues

3.6.2.1 Start-Up

Startup deals with the settlement of an adequate population of microbes inside the ATU to digest the nitrogenous components and soluble organic of the influent.Utmost applications initiate the process by entering an ample population of microbes with the wastewater to the ATU.From an operational ATU, insertion of a certain gallon of the mixed liquor is one of the methods of inoculating the system if required (Rozich & Bordacs, 2002; Kassab et al., 2010).The microbes seem to be scattered and flocs have not been formed that will be considered to be settled in the clarifier during the increment in biomass concentration.Solids carry over has the leading potential exclusively with high hydraulic loads, as far as the biomass can settle more quickly and becomes more flocculated.If there is a permission for solids to boost up in the clarifier, gas will produce in the biosolids due to anaerobic conditions within the solids and make solids to lift to the surface and create a scum layer.The judgment of the quality of the triggered sludge is a valuable measure of how healthy the process is continuing.Normally, high quality activated sludge, if kept aerated, will have an earthy smell, and a golden-brown color (Leitão et al., 2006).

3.6.2.2 Typical Problems

On hiking of filamentous bacteria (chiefly *sphaerotilus)* that is attached to the floc particles and obstruct settling is a cause of the phenomenon developed in the aeration tank known as sludge bulking (Crites and Tchobanoglous, 1998).Vast changes can be tolerated by such microorganisms in nutrients and dissolved oxygen, often situations occurring in low-scale aerobic treatment units.Carryover of solids remains in the effluent on the occurrence of these conditions.This phenomenon is bothersome especially those small-scale plants that have a shortage of technical support and notable inconstancy in organic loading (Fricke et al., 2007; Chan et al.,2012).

In the aeration chamber, frothing and foaming on the liquid surface (and the clarifier) results in an occurrence of superfluous growth of *nocardia* (a hydrophobic bacterium).The fact makes the problem worsen that more growth of the foster and foam is trapped by the baffles in the clarifier trap (Crites and Tchobanoglous, 1998). The froth spray pumps are supplied by some of the manufacturers of ATU.The froth spray helps to decompose the froth and cut down the surface tension of the water (OhioEPA, 2000).

3.6.2.3 Biomass (Sludge) Wastage

Although the extended aeration process is used by ATUs, aggregation of old biomass cannot be completely prevented by endogenous degradation. Non-biodegradable solids and biomass will gather at the bottom of the ATU, and systematically, a portion of these solids must be dismissed by a maintenance provider.While removing, some of the solids are essentially left in the aerobic chamber to assist as the seed to breed the biological floc (Maranon et al., 2008).

3.6.3 Performance Certification

A standardized method is published by NSF International and the American National Standards Institute for autonomous analysts to certify the reliability and performance of aeration units. NSF/ANSI Standard 40-2000 which is Residential Wastewater Treatment Systems, build design and construction, least materials, and performance demands for residential wastewater treatment systems having treatment space between 400 and 1500 gallons per day and defined, single discharge points (Kheradmand et al., 2010).

3.6.3.1 Mechanical Evaluation

Construction and design concerns of this standard make sure that during the subjection of the system to earth and hydrostatic pressures, the structural integrity maintains.During and later the performance testing cycle, an *in situ* visual assessment of the structural elements is done.To ensure that the system is watertight, it is tested (i.e., exfiltration of wastewater or no infiltration of groundwater).Watertightness is figured out by pouring the tank with tap water to the stage of the high-level buzzer. Then 24 hours monitoring is done for this level.

All ATUs have mobile units.These parts work in very destructive environments and thus demands regular maintenance or/and substitute.All mechanical parts are checked out at the time of certification procedure to figure out the regularity of required maintenance and the comfort of the service provider for maintenance performance.For the assurance of the protection of all electrical components by safety devices that meet or beat Standard ANSI/NFPA 70, audits are conducted.There must have processes or mechanisms for ATUs that can be able to detect the collapse of those mechanical and electrical components that are discovering high water situations or/and integral to the treatment processes.These techniques must be proficient in notifying the owner on the occurrence of mechanical, electrical, or hydraulic malfunction by audible and visible signal (Chan et al., 2009).For periodic cleaning, visual inspection, replacement of components, sampling, and removal of residuals, there must have ground-level access ports by all units.Ports access must be shielded against illegal intrusions via covers compelling the use of special gadgets, padlocks, or a cover measuring a minimum of 65 pounds.

3.6.3.2 Performance Evaluation

Performance evaluation and testing of a particular model/kind of treatment system being operated for 26 successive weeks from which 16 weeks are utilized for design loading pursued by stress loading (7.5 weeks) and the remaining 2.5 weeks of design loading.Design loading which involves the operation of 7 days per week with a volume of wastewater, is equivalent to the day-to-day hydraulic space of the unit.The wastewater's total suspended solids (TSS) concentrations getting into the system should range between 100-350 mg/L, and 30-day average carbonaceous BOD5 (CBOD5) must range between 100-300 mg/L.Four non-design conditions are simulated to design stress loading i.e. (i) laundry day, (ii) working parents, (iii) equipment or power failure, and (iv) vacation (Sharma & Singh, 2013; Chernicharo et al., 2015).

Non-interruptive performance evaluation and testing for maintenance or routine service are regulated all along 96 data days.Else except specified, analysis methods and all sample compilation must be according to the current issue of the American Public Health Association's Standard Methods for the Examination of Water and Wastewater.Composite effluent samples are gathered and analyzed regularly 5 days per week within a duration of design loading.During the conditions of stress loading, on the day every stress condition is started, effluent and influent 24-hr composite samples are acquired.Later, to observe the resumption of the treatment unit, samples are captured.24 hours after finalization of the wash day, vacation stresses, and working-parent, 24 hours effluent and influent composite samples are gathered for consecutive 6 days.48 hours after the completion of the equipment/power failure stress, effluent, and influent 24 hours of composite samples are captured for consecutive 5 days (Darwati & Hastuti, 2019).

The categorization of residential wastewater treatment systems is based on the biological, physical, and chemical features of their effluents either Class I or Class II.A Class I certification demonstrates performance for three parameters – pH, CBOD5, and solids to EPA Secondary Treatment

Guidelines (U.S. EPA, 1996).A unit is tolerated to surpass 1.4 times the effluent extent for TSS sample concentrations and CBOD5 without dooming status of Class I for the first calendar month of performance evaluation and testing.If 10 percent or less of their effluent CBOD5 is greater than 60 mg/L and TSS sample concentrations are greater than 100 mg/L, then a system can be labeled with Class II.The benchmark for Class I and Class II performance standards are provided in Table 3.2 (Seabloom & Buchanan, 2005; Sharma & Singh, 2013).

Table 3.2: Performance classifications of NSF/ANSI Standard Number 40-2000

Class I	**Parameter**	CBOD$_5$	TSS	Color	Threshold Odor	Oily Film	Foam	pH
	30 days average shall not exceed	25 mg/L	30 mg/L	Each sample shall be less than 15 NTU units.	Non-offensive	None seeable other than air bubbles	None	Every effluent sample shall be between 6.0 and 9.0
	7 days average shall not exceed	40 mg/L	45 mg/L					
Class II	Less than 10% of the effluent BOD$_5$ values shall surpass 60 mg/L and less than 10% of the effluent TSS values shall surpass 100 mg/L.							

The performance bases of NSF/ANSI Standard 40 are suspended solids in the effluent and organic carbon shown in Table 3.2. Evaluating treatment units have an increased interest in their scope to remove phosphorus, pathogens, and nitrogen (Maxfield et al., 2003; Williamson, 2010). The methods for the evaluation of the evacuation of these elements are provided by Standard 40. However, for receiving certification, there is no requirement of specific performance of the eviction of these constituents (Converse, 2001). Removal of solids and digestion of colloidal and soluble organic compounds is the primary characteristic of saturated, aerobic treatment units. To provide phosphorus removal, disinfection, and denitrification extra unit processes are combined with the treatment series (Eley & Morehouse, 2003; Moelants et al., 2008).

REFERENCES

1. Abdalla, K. Z., & Hammam, G. (2014). Correlation between biochemical oxygen demand and chemical oxygen demand for various wastewater treatment plants in Egypt to obtain the biodegradability indices. *International Journal of Sciences: Basic and Applied Research*, *13*(1), 42-48.
2. Achilli, A., Marchand, E. A., & Childress, A. E. (2011). A performance evaluation of three membrane bioreactor systems: aerobic, anaerobic, and attached-growth. *Water Science and Technology*, *63*(12), 2999-3005.
3. Aivasidis, A., & Diamantis, V. I. (2005). Biochemical reaction engineering and process development in anaerobic wastewater treatment. In *Technology Transfer in Biotechnology* (Vol.1, pp. 49-76). Springer, Berlin, Heidelberg.
4. Alvarino, T., Suárez, S., Garrido, M., Lema, J. M., & Omil, F. (2016). A UASB reactor coupled to a hybrid aerobic MBR as innovative plant configuration to enhance the removal of organic micropollutants. *Chemosphere*, *144*, 452-458.
5. Amuda, O. S., & Ibrahim, A. O. (2006). Industrial wastewater treatment using natural material as adsorbent. *African Journal of Biotechnology*, *5*(16), 1-33.
6. Anagnostopoulos, K., & Vavatsikos, A. (2012). Site suitability analysis for natural systems for wastewater treatment with spatial fuzzy analytic hierarchy process. *Journal of Water Resources Planning and Management*, *138*(2), 125-134.
7. Arya, V., Philip, L., & Bhallamudi, S. M. (2016). Performance of suspended and attached growth bioreactors for the removal of cationic and anionic pharmaceuticals. *Chemical Engineering Journal*, *284*, 1295-1307.
8. ASCE, 1977. Wastewater Treatment Plant Design, Manual of Practice No. 36, pp.1-44. Lancaster Press, Lancaster, PA.
9. Aura, A. M. (2008). Microbial metabolism of dietary phenolic compounds in the colon. *Phytochemistry Reviews*, *7*(3), 407-429.
10. Béline, F., Boursier, H., Daumer, M. L., Guiziou, F., & Paul, E. (2007). Modelling of biological processes during aerobic treatment of piggery wastewater aiming at process optimisation. *Bioresource technology*, *98*(17), 3298-3308.

11. Benefield, L. D. and C. W. Randall. 1985. Biological Process Design for Waste Water Treatment (Vol.1, pp. 1-33). Ibis Publishing, Charlottesville, Virginia.
12. Bertanza, G. (1997). Simultaneous nitrification-denitrification process in extended aeration plants: pilot and real scale experiences. *Water Science and Technology*, *35*(6), 53.
13. Blair, N. E. A. L., Leu, A., Muñoz, E. L. A. I. N. E., Olsen, J. E. R. R. Y., Kwong, E. M. I. L., & Des Marais, D. A. V. I. D. (1985). Carbon isotopic fractionation in heterotrophic microbial metabolism. *Appl. Environ. Microbiol.*, *50*(4), 996-1001.
14. Bounds, T. 2003. Personal correspondence (Vol.1, pp. 1-44). Orenco Systems, Inc. May 24.
15. Cao, Y., van Loosdrecht, M. C., & Daigger, G. T. (2017). Mainstream partial nitritation–anammox in municipal wastewater treatment: status, bottlenecks, and further studies. *Applied microbiology and biotechnology*, *101*(4), 1365-1383.
16. Cema, G., Wiszniowski, J., Żabczyński, S., Zabłocka-Godlewska, E., Raszka, A., & Surmacz-Górska, J. (2007). Biological nitrogen removal from landfill leachate by deammonification assisted by heterotrophic denitrification in a rotating biological contactor (RBC). *Water Science and Technology*, *55*(8-9), 35-42.
17. Cerniglia, C. E. (1984). Microbial metabolism of polycyclic aromatic hydrocarbons. In *Advances in applied microbiology* (Vol. 30, pp. 31-71). Academic Press.
18. Chan, Y. J., Chong, M. F., & Law, C. L. (2012). An integrated anaerobic–aerobic bioreactor (IAAB) for the treatment of palm oil mill effluent (POME): Start-up and steady state performance. *Process Biochemistry*, *47*(3), 485-495.
19. Chan, Y. J., Chong, M. F., Law, C. L., & Hassell, D. G. (2009). A review on anaerobic–aerobic treatment of industrial and municipal wastewater. *Chemical Engineering Journal*, *155*(1-2), 1-18.
20. Chan, Y. J., Chong, M. F., Law, C. L., & Hassell, D. G. (2009). A review on anaerobic–aerobic treatment of industrial and municipal wastewater. *Chemical Engineering Journal*, *155*(1-2), 1-18.
21. Chernicharo, C. A. L., Van Lier, J. B., Noyola, A., & Ribeiro, T. B. (2015). Anaerobic sewage treatment: state of the art, constraints and challenges. *Reviews in Environmental Science and Bio/*

Technology, *14*(4), 649-679.

22. Chipasa, K. B. (2003). Accumulation and fate of selected heavy metals in a biological wastewater treatment system. *Waste management*, *23*(2), 135-143.
23. Chubukov, V., Gerosa, L., Kochanowski, K., & Sauer, U. (2014). Coordination of microbial metabolism. *Nature Reviews Microbiology*, *12*(5), 327-340.
24. Comeau, Y., Hall, K. J., Hancock, R. E. W., & Oldham, W. K. (1986). Biochemical model for enhanced biological phosphorus removal. *Water Research*, *20*(12), 1511-1521.
25. Converse, 2001. Aeration treatment of onsite domestic wastewater, aerobic units and packed bed filters (Vol.1, pp. 1-41). Small Scale Waste Management Project, University of Wisconsin, Madison, www.wisc.edu/sswmp. Publication List.
26. Crites, R. and G. Tchobanoglous. 1998. Small and Decentralized Wastewater Management (Vol. 1, pp. 1-44). McGraw-Hill, Boston.
27. Darwati, S., & Hastuti, E. (2019, March). Developing of Performance Testing Scheme for Decentralized Domestic Wastewater Treatment Plant in Indonesia. In *Proceeding International Conference on Science and Engineering* (Vol. 2, pp. 301-307).
28. Davis, M. L. and D. A. Cornwell. 1991. Introduction to Environmental Engineering (Vol.1, pp. 1-53). McGraw-Hill, New York.
29. Delmer, D. P. (1987). Cellulose biosynthesis. *Annual review of plant physiology*, *38*(1), 259-290.
30. Delmer, D. P., & Amor, Y. (1995). Cellulose biosynthesis. *The plant cell*, *7*(7), 987.
31. Dharmsthiti, S., & Kuhasuntisuk, B. (1998). Lipase from Pseudomonas aeruginosa LP602: biochemical properties and application for wastewater treatment. *Journal of Industrial Microbiology and Biotechnology*, *21*(1-2), 75-80.
32. Dogan, E., Ates, A., Yilmaz, E. C., & Eren, B. (2008). Application of artificial neural networks to estimate wastewater treatment plant inlet biochemical oxygen demand. *Environmental progress*, *27*(4), 439-446.
33. Egami, F. (1973). A comment to the concept on the role of nitrate fermentation and nitrate respiration in an evolutionary pathway of energy metabolism. *Zeitschrift für allgemeine Mikrobiologie*, *13*(2), 177-181.

34. Eikum, A. and T. Benett. 1992. New Norwegian technology for treatment of small flows. Proceedings of the seventh Northwest on-site Wastewater Treatment Short Course and Equipment Exhibition (Vol.1, pp. 1-23). University of Washington, Seattle, WA.

35. Eley, W. D., & Morehouse, C. H. (2003, September). Evaluation of new technology for shipboard wastewater treatment. In *Oceans 2003. Celebrating the Past... Teaming Toward the Future (IEEE Cat. No. 03CH37492)* (Vol. 2, pp. 748-753). IEEE.

36. Fan, X. J., Urbain, V., Qian, Y., & Manem, J. (1996). Nitrification and mass balance with a membrane bioreactor for municipal wastewater treatment. *Water Science and Technology*, *34*(1-2), 129-136.

37. Freitas, T. K. F. S., Oliveira, V. M., De Souza, M. T. F., Geraldino, H. C. L., Almeida, V. C., Fávaro, S. L., & Garcia, J. C. (2015). Optimization of coagulation-flocculation process for treatment of industrial textile wastewater using okra (A. esculentus) mucilage as natural coagulant. *Industrial Crops and Products*, *76*, 538-544.

38. Fricke, K., Santen, H., Wallmann, R., Hüttner, A., & Dichtl, N. (2007). Operating problems in anaerobic digestion plants resulting from nitrogen in MSW. *Waste Management*, *27*(1), 30-43.

39. Fujie, K., Hu, H. Y., Ikeda, Y., & Urano, K. (1992). Gas-liquid oxygen transfer characteristics in an aerobic submerged biofilter for the wastewater treatment. *Chemical engineering science*, *47*(13-14), 3745-3752.

40. Gavrilescu, M., & Macoveanu, M. (1999). Process engineering in biological aerobic waste□water treatment. *Acta biotechnologica*, *19*(2), 111-145.

41. Gouveia, J., Plaza, F., Garralon, G., Fdz-Polanco, F., & Peña, M. (2015). A novel configuration for an anaerobic submerged membrane bioreactor (AnSMBR). Long-term treatment of municipal wastewater under psychrophilic conditions. *Bioresource technology*, *198*, 510-519.

42. Heddam, S., Lamda, H., & Filali, S. (2016). Predicting effluent biochemical oxygen demand in a wastewater treatment plant using generalized regression neural network based approach: a comparative study. *Environmental Processes*, *3*(1), 153-165.

43. Heger, S. F., Schmidt, D. R., & Janni, K. A. (2010). Aerobic and media filter treatment systems for milk house wastewater on small dairy operations. *Applied engineering in agriculture*, *26*(2), 319-327.

44. Henry, J.G. and G. Heinke. 1996. Environmental Science and Engineering (Vol. 2, pp. 1-55). Prentice Hall, Upper Saddle River, N.J.
45. Holler, S., & Trösch, W. (2001). Treatment of urban wastewater in a membrane bioreactor at high organic loading rates. *Journal of biotechnology*, *92*(2), 95-101.
46. Ichinari, T., Ohtsubo, A., Ozawa, T., Hasegawa, K., Teduka, K., Oguchi, T., & Kiso, Y. (2008). Wastewater treatment performance and sludge reduction properties of a household wastewater treatment system combined with an aerobic sludge digestion unit. *Process Biochemistry*, *43*(7), 722-728.
47. Irvine, R. L., Ketchum Jr, L. H., & Asano, T. (1989). Sequencing batch reactors for biological wastewater treatment. *Critical Reviews in Environmental Science and Technology*, *18*(4), 255-294.
48. Ivery, G. (1996). Aerobic treatment units (ATUs): appropriate technology for on-site wastewater treatment and re-use. *Desalination*, *106*(1-3), 295-303.
49. Iyo, T., Yoshino, T., Tadokoro, M., Ogawa, T., & Ohno, S. (1996a). Advanced performance of small-scale domestic sewage treatment plants using anaerobic-aerobic filter systems with flow-equalization and recirculation. *Environmental technology*, *17*(11), 1235-1243.
50. Iyo, T., Yoshino, T., Tadokoro, M., OGAWA, T., OHNO, S., & SEKI, Y. (1996b). Simultaneous Removal of BOD, Nitrogen, and Phosphorus in Small-Scale Domestic Sewage Treatment Plant using Anaerobic-Aerobic Filter Systems with Flow-Equalization, Recirculation, and Aluminum Biofilter Media. *Japanese Journal of Water Treatment Biology*, *32*(2), 131-142.
51. Jern, N. W. (1987). Aerobic treatment of piggery wastewater with the sequencing batch reactor. *Biological wastes*, *22*(4), 285-294.
52. Jianmin, K. Q. M. W. X., & Fangwei, L. G. F. (2005). Effect of hyper-saline concentration on the biochemical treatment of industrial wastewater [J]. *Techniques and Equipment for Environmental Pollution Control*, *8, 1-44*.
53. Joss, A., Keller, E., Alder, A. C., Göbel, A., McArdell, C. S., Ternes, T., & Siegrist, H. (2005). Removal of pharmaceuticals and fragrances in biological wastewater treatment. *Water research*, *39*(14), 3139-3152.
54. Kassab, G., Halalsheh, M., Klapwijk, A., Fayyad, M., & Van Lier, J. B. (2010). Sequential anaerobic–aerobic treatment for domestic

wastewater–A review. *Bioresource Technology*, *101*(10), 3299-3310.

55. Kassab, G., Halalsheh, M., Klapwijk, A., Fayyad, M., & Van Lier, J. B. (2010). Sequential anaerobic–aerobic treatment for domestic wastewater–A review. *Bioresource Technology*, *101*(10), 3299-3310.
56. Kende, H. (1993). Ethylene biosynthesis. *Annual review of plant biology*, *44*(1), 283-307.
57. Kerner, M. (1993). Coupling of microbial fermentation and respiration processes in an intertidal mudflat of the Elbe estuary. *Limnology and Oceanography*, *38*(2), 314-330.
58. Kheradmand, S., Karimi-Jashni, A., & Sartaj, M. (2010). Treatment of municipal landfill leachate using a combined anaerobic digester and activated sludge system. *Waste Management*, *30*(6), 1025-1031.
59. Kumar, T., Bhargava, R., Prasad, K. H., & Pruthi, V. (2015). Evaluation of vermifiltration process using natural ingredients for effective wastewater treatment. *Ecological engineering*, *75*, 370-377.
60. La Motta, E. J., Silva, E., Bustillos, A., Padrón, H., & Luque, J. (2007). Combined anaerobic/aerobic secondary municipal wastewater treatment: pilot-plant demonstration of the UASB/aerobic solids contact system. *Journal of environmental engineering*, *133*(4), 397-403.
61. Larson, R. A., & Safferman, S. I. (2009). Aerobic Treatment Unit Performance on Milking Parlor Wash Water. *Transactions of the ASABE*, *52*(3), 825-834.
62. Larson, R. A., & Safferman, S. I. (2009). Aerobic Treatment Unit Performance on Milking Parlor Wash Water. *Transactions of the ASABE*, *52*(3), 825-834.
63. Lee, J., Ahn, W. Y., & Lee, C. H. (2001). Comparison of the filtration characteristics between attached and suspended growth microorganisms in submerged membrane bioreactor. *Water Research*, *35*(10), 2435-2445.
64. Lee, J., Ahn, W. Y., & Lee, C. H. (2001). Comparison of the filtration characteristics between attached and suspended growth microorganisms in submerged membrane bioreactor. *Water Research*, *35*(10), 2435-2445.
65. Lee, N. M., & Welander, T. (1996). Reducing sludge production in aerobic wastewater treatment through manipulation of the ecosystem. *Water Research*, *30*(8), 1781-1790.

66. Lehninger, A. L. 1973. Bioenergetics, Vol.2, pp. 1-44. W. A. Benjamin, Inc. Menlo Park, California.

67. Leitão, R. C., Van Haandel, A. C., Zeeman, G., & Lettinga, G. (2006). The effects of operational and environmental variations on anaerobic wastewater treatment systems: a review. *Bioresource Technology*, *97*(9), 1105-1118.

68. Lo, C. K., Yu, C. W., Tam, N. F. Y., & Traynor, S. (1994). Enhanced nutrient removal by oxidation-reduction potential (ORP) controlled aeration in a laboratory scale extended aeration treatment system. *Water Research*, *28*(10), 2087-2094.

69. Makinen, P. M., Theno, T. J., Ferguson, J. F., Ongerth, J. E., & Puhakka, J. A. (1993). Chlorophenol toxicity removal and monitoring in aerobic treatment: recovery from process upsets. *Environmental science & technology*, *27*(7), 1434-1439.

70. Maranon, E., Vazquez, I., Rodriguez, J., Castrillon, L., Fernandez, Y., & Lopez, H. (2008). Treatment of coke wastewater in a sequential batch reactor (SBR) at pilot plant scale. *Bioresource technology*, *99*(10), 4192-4198.

71. Martin, E. and T. Martin. 1991. "Technologies for small water and wastewater systems (Vol. 1, pp. 1-33). Van Nostrand Reinhold, NY.

72. Maxfield, M., Daniell, W. E., Treser, C. D., & VanDerslice, J. (2003). Aerobic residential onsite sewage systems: An evaluation of treated-effluent quality. *Journal of environmental health*, *66*(3), 14.

73. Metcalf and Eddy, Inc. 2003. Wastewater Engineering: Treatment and Reuse (Vol.4, pp. 1-52). McGraw-Hill, Boston.

74. Middlebrooks, E. J., & Garland, C. F. (1968). Kinetics of model and field extended-aeration wastewater treatment units. *Journal (Water Pollution Control Federation)*, 1, 586-612.

75. Moelants, N., Janssen, G., Smets, I., & Van Impe, J. (2008). Field performance assessment of onsite individual wastewater treatment systems. *Water Science and Technology*, *58*(1), 1-6.

76. Mohan, S. V., Raghavulu, S. V., & Sarma, P. N. (2008). Biochemical evaluation of bioelectricity production process from anaerobic wastewater treatment in a single chambered microbial fuel cell (MFC) employing glass wool membrane. *Biosensors and Bioelectronics*, *23*(9), 1326-1332.

77. Mohan, S. V., Rao, N. C., Prasad, K. K., Madhavi, B. T. V., &

Sharma, P. N. (2005). Treatment of complex chemical wastewater in a sequencing batch reactor (SBR) with an aerobic suspended growth configuration. *Process Biochemistry*, *40*(5), 1501-1508.

78. Moosvi, S., & Madamwar, D. (2007). An integrated process for the treatment of CETP wastewater using coagulation, anaerobic and aerobic process. *Bioresource technology*, *98*(17), 3384-3392.
79. Muñoz, R., Díaz, L. F., Bordel, S., & Villaverde, S. (2008). Response of Pseudomonas putida F1 cultures to fluctuating toluene loads and operational failures in suspended growth bioreactors. *Biodegradation*, *19*(6), 897-908.
80. Najafpour, G. D., Zinatizadeh, A. A. L., & Lee, L. K. (2006). Performance of a three-stage aerobic RBC reactor in food canning wastewater treatment. *Biochemical engineering journal*, *30*(3), 297-302.
81. OhioEPA. 2000. Guide for owners of package extended aeration sewage treatment plants, operation and maintenance (Vol.1, pp. 1-33). State of Ohio EPA, Division of Surface Water.
82. Ouyang, X., Guo, F., Shan, D., Yu, H., & Wang, J. (2015). Development of the integrated fuzzy analytical hierarchy process with multidimensional scaling in selection of natural wastewater treatment alternatives. *Ecological Engineering*, *74*, 438-447.
83. Passoth, V., Zimmermann, M., & Klinner, U. (1996). Peculiarities of the regulation of fermentation and respiration in the crabtree-negative, xylose-fermenting yeastPichia stipitis. *Applied biochemistry and biotechnology*, *57*(1), 201.
84. Perle, M., Kimchie, S., & Shelef, G. (1995). Some biochemical aspects of the anaerobic degradation of dairy wastewater. *Water Research*, *29*(6), 1549-1554.
85. Pomiès, M., Choubert, J. M., Wisniewski, C., & Coquery, M. (2013). Modelling of micropollutant removal in biological wastewater treatments: a review. *Science of the Total Environment*, *443*, 733-748.
86. Pophali, G. R., Khan, R., Dhodapkar, R. S., Nandy, T., & Devotta, S. (2007). Anaerobic–aerobic treatment of purified terephthalic acid (PTA) effluent; a techno-economic alternative to two-stage aerobic process. *Journal of environmental management*, *85*(4), 1024-1033.
87. Pynaert, K., Smets, B. F., Wyffels, S., Beheydt, D., Siciliano, S. D., & Verstraete, W. (2003). Characterization of an autotrophic nitrogen-

removing biofilm from a highly loaded lab-scale rotating biological contactor. *Appl. Environ. Microbiol.*, *69*(6), 3626-3635.

88. Reynolds, T. D. 1982. Unit Operations and Processes in Environmental Engineering (Vol.1, pp. 1-22). PWS-KENT Publishing Company, Boston.
89. Rodríguez, F. A., Poyatos, J. M., Reboleiro-Rivas, P., Osorio, F., González-López, J., & Hontoria, E. (2011). Kinetic study and oxygen transfer efficiency evaluation using respirometric methods in a submerged membrane bioreactor using pure oxygen to supply the aerobic conditions. *Bioresource technology*, *102*(10), 6013-6018.
90. Roeder, E., & Brookman, W. (2006). Performance of aerobic treatment units: monitoring results from the Florida Keys. *Journal of environmental health*, *69*(4), 1-33.
91. Rozich, A. F., & Bordacs, K. (2002). Use of thermophilic biological aerobic technology for industrial waste treatment. *Water science and technology*, *46*(4-5), 83-89.
92. S. EPA. 1992. Summary Report: "Small community water and wastewater treatment. Office of Water, Washington, D. C., EPA/625/R-92/010,pp. 1-55
93. Schaar, H., Clara, M., Gans, O., & Kreuzinger, N. (2010). Micropollutant removal during biological wastewater treatment and a subsequent ozonation step. *Environmental Pollution*, *158*(5), 1399-1404.
94. Schuetz, R., Zamboni, N., Zampieri, M., Heinemann, M., & Sauer, U. (2012). Multidimensional optimality of microbial metabolism. *Science*, *336*(6081), 601-604.
95. Schwarzenbeck, N., Erley, R., & Wilderer, P. A. (2004). Aerobic granular sludge in an SBR-system treating wastewater rich in particulate matter. *Water Science and Technology*, *49*(11-12), 41-46.
96. Seabloom, R. W., & Buchanan, J. R. (2005). Aerobic Treatment Units–PowerPoint Presentation. in (MA Gross and NE Deal, eds.) University Curriculum Development for Decentralized Wastewater Management. *National Decentralized Water Resources Capacity Development Project (vol.1, pp. 1-33). University of Arkansas, Fayetteville, AR.*
97. Sharma, C., & Singh, S. K. (2013). Performance Evaluation Of Sewage Treatment Plant Based On Advanced Aerobic Biological Filtration And Oxygenated Reactor (BIOFOR) Technology-A Case Study Of Capital

City-Delhi, India. *International Journal of Engineering Science and Innovative Technology*, *2*(4), 1-30.

98. Sharma, C., & Singh, S. K. (2013). Performance Evaluation Of Sewage Treatment Plant Based On Advanced Aerobic Biological Filtration And Oxygenated Reactor (BIOFOR) Technology-A Case Study Of Capital City-Delhi, India. *International Journal of Engineering Science and Innovative Technology*, *2*(4), 1-33.
99. Sombatsompop, K., Visvanathan, C., & Aim, R. B. (2006). Evaluation of biofouling phenomenon in suspended and attached growth membrane bioreactor systems. *Desalination*, *201*(1-3), 138-149.
100. Sombatsompop, K., Visvanathan, C., & Aim, R. B. (2006). Evaluation of biofouling phenomenon in suspended and attached growth membrane bioreactor systems. *Desalination*, *201*(1-3), 138-149.
101. Stewart, M. J., Ludwig, H. F., & Kearns, W. H. (1962). Effects of varying salinity on the extended aeration process. *Journal (Water Pollution Control Federation)*, *34*(11), 1161-1177.
102. Stolz, J. F., Basu, P., Santini, J. M., & Oremland, R. S. (2006). Arsenic and selenium in microbial metabolism. *Annu. Rev. Microbiol.*, *60*, 107-130.
103. Sui, Q., Huang, J., Deng, S., Chen, W., & Yu, G. (2011). Seasonal variation in the occurrence and removal of pharmaceuticals and personal care products in different biological wastewater treatment processes. *Environmental science & technology*, *45*(8), 3341-3348.
104. Suzuki, S., Karube, I., & Matsunaga, T. (1978, January). Application of a biochemical fuel cell to wastewaters. In *Biotechnol. Bioeng. Symp.;(United States)* (Vol. 8, pp. 1-44). Tokyo Inst. of Tech., Yokohama, Japan.
105. Tang, W. W., Wang, Z., Levison, B. S., Koeth, R. A., Britt, E. B., Fu, X., ... & Hazen, S. L. (2013). Intestinal microbial metabolism of phosphatidylcholine and cardiovascular risk. *New England Journal of Medicine*, *368*(17), 1575-1584.
106. Tanner, C. C., Sukias, J. P., Headley, T. R., Yates, C. R., & Stott, R. (2012). Constructed wetlands and denitrifying bioreactors for on-site and decentralised wastewater treatment: comparison of five alternative configurations. *Ecological Engineering*, *42*, 112-123.
107. Taylor, D. L. (1942). Influence of oxygen tension on respiration, fermentation, and growth in wheat and rice. *American Journal of*

Botany, 1, 721-738.

108. Thanh, B. X., Visvanathan, C., Spérandio, M., & Aim, R. B. (2008). Fouling characterization in aerobic granulation coupled baffled membrane separation unit. *Journal of Membrane Science*, *318*(1-2), 334-339.

109. Tziotzios, G., Teliou, M., Kaltsouni, V., Lyberatos, G., & Vayenas, D. V. (2005). Biological phenol removal using suspended growth and packed bed reactors. *Biochemical engineering journal*, *26*(1), 65-71.

110. U. S. EPA. 1986. Summary Report: Sequencing Batch Reactors. Office of Water, Washington, D.C., EPA 625886011, pp. 1-100

111. U. S. EPA. 2000. Decentralized Systems Technology Fact Sheet - Aerobic Treatment. Office of Water, Washington, D.C., EPA 832-F-00-031, pp. 1-44.

112. U. S. EPA. 2002. Onsite Wastewater Treatment Systems Manual. Office of Water, Washington, D. C., EPA/625/R-00/008, pp. 1-55.

113. U.S. EPA, 1996 Permit Writers' Manual, Office of Water, Washington, D. C., EPA-833B-96-003.

114. Van Lier, J. B., & Boncz, M. A. (2002). Controlling calcium precipitation in an integrated anaerobic–aerobic treatment system of a "zero-discharge" paper mill. *Water science and technology*, *45*(10), 341-347.

115. Verduyn, C., Postma, E., Scheffers, W. A., & Van Dijken, J. P. (1992). Effect of benzoic acid on metabolic fluxes in yeasts: a continuous□ culture study on the regulation of respiration and alcoholic fermentation. *Yeast*, *8*(7), 501-517.

116. Vogt, T. (2010). Phenylpropanoid biosynthesis. *Molecular plant*, *3*(1), 2-20.

117. Von Wettstein, D., Gough, S., & Kannangara, C. G. (1995). Chlorophyll biosynthesis. *The Plant Cell*, *7*(7), 1039.

118. Wang, J. P., Chen, Y. Z., Wang, Y., Yuan, S. J., & Yu, H. Q. (2011). Optimization of the coagulation-flocculation process for pulp mill wastewater treatment using a combination of uniform design and response surface methodology. *Water research*, *45*(17), 5633-5640.

119. Wang, S., & Peng, Y. (2010). Natural zeolites as effective adsorbents in water and wastewater treatment. *Chemical Engineering Journal*, *156*(1), 11-24.

120. Wang, W., Jung, Y. J., Kiso, Y., Yamada, T., & Min, K. S. (2006). Excess

sludge reduction performance of an aerobic SBR process equipped with a submerged mesh filter unit. *Process biochemistry*, *41*(4), 745-751.

121. Wei, Y., Van Houten, R. T., Borger, A. R., Eikelboom, D. H., & Fan, Y. (2003). Minimization of excess sludge production for biological wastewater treatment. *Water Research*, *37*(18), 4453-4467.
122. Westerhoff, P. K., Kiser, M. A., & Hristovski, K. (2013). Nanomaterial removal and transformation during biological wastewater treatment. *Environmental engineering science*, *30*(3), 109-117.
123. Williamson, E. (2010). Cold Climate Performance Analysis of On□ Site Domestic Wastewater Treatment Systems. *Water environment research*, *82*(6), 512-518.
124. Zhang, T., Bu, P., Zeng, J., & Vancura, A. (2017). Increased heme synthesis in yeast induces a metabolic switch from fermentation to respiration even under conditions of glucose repression. *Journal of Biological Chemistry*, *292*(41), 16942-16954.
125. Zhang, Y., Pinoy, L., Meesschaert, B., & Van der Bruggen, B. (2013). A natural driven membrane process for brackish and wastewater treatment: photovoltaic powered ED and FO hybrid system. *Environmental science & technology*, *47*(18), 10548-10555.

CHAPTER

4

ACTIVATED SLUDGE PROCESS

CONTENTS

4.1 INTRODUCTION

Lately, there has been a consistent rise in the laws related to the purification practices required of the wastewater treatment plants by the administrative agencies responsible for managing sewerage. Due to this, technical difficulties along with the number of wastewater treatment plants have been increasing. To control the operation and capital expenses within limits, developmental process technology solutions need to be formed (Shaul et al., 1991; Liu &Tay, 2001). Usually, treatment plants include gas utilization, biological and mechanical cleansing of water, and sludge treatment. In addition to analyzing the individual procedures, it is necessary to review the system as a whole. The procedures used for the separation of solids/liquids are extremely essential in three of these four stages .i.e. the initial and secondary purification of the wastewater and in the thickening and drying of the sludge. The characteristics of the sludge greatly influence the proficiency of the solids/liquids separation (Yasui& Shibata, 1994; Liu, 2003).

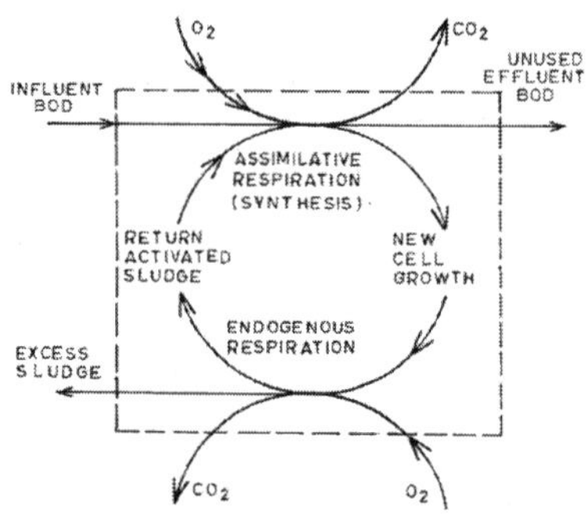

Figure 4.1: The activated sludge process's reaction system.

[Source: http://www.engineeringenotes.com/waste-management/activated-sludge-process/activated-sludge-process-of-sewage-treatment-waste-management/40150]

The Activated Sludge (AS) process is favorable for a large population where the property is premium, it was expanded as an alternative for biological filters. This system can be operated in several different conditions, as highlighted by numerous latest researches, constructing it a more workable technique than biological filtration (Adham et al., 1996; Gros et al., 2010). The Activated Sludge process happens to bean applied natural biological treatment approach. It is an intricate combination of biochemistry and microbiology introducing various kinds of defects. The bacteria in the Activated Sludge Plant (ASP) release adhesive matter that cloaks the microscopic particles present in the wastewater, creating flocks of a gel-like material, setting up support on, and in which, and the bugs are present. The activated sludge is chocolate-brown (Balkema et al.,1998; Hernández-Esparza et al., 2006). The bugs can consume the organic matter (BOD) through the aeration of the activated sludge to dissolve oxygen. The food or the organic substance gets integrated with the activated sludge. In addition to providing the bugs the ability to utilize food (BOD) the dissolved oxygen also transforms ammonia to nitrate. For the chemical changes to occur properly, the tank should be large and adequate to allow significant contact time (retention time) between the activated sludge and wastewater (American Water Works Association, 1990; Ahansazan et al., 2014).

A technique for the biological treatment of wastewater, Activated Sludge Treatment incorporates Activated sludge into wastewater which accelerates the breakdown of waste. The solution is then churned up and exposed to air for a particular amount of time ultimately causing the activated sludge to sediment. Finally, it is either recycled (sent back to the aeration tank) or disposed of (wasted). The activated sludge process is capable of manufacturing high-quality waste within an acceptable maintenance and functioning cost. Handling waste economically is beneficial for everyone in society. Microbes consume the organic adulterants present in the discharge in the activated sludge process, forming high-quality waste. The fundamental principle behind the activated sludge processes is that as the microorganisms grow, they tend to form particles that adhere together; this floc is then allowed to settle down at the base of the tank, leaving a fairly clear liquid uncontaminated by organic material and suspended solids. This chapter aims to outline and describe the guidelines related to the Activated Sludge Treatment Process (Nagaoka et al., 1996; Sakai et al., 1997).

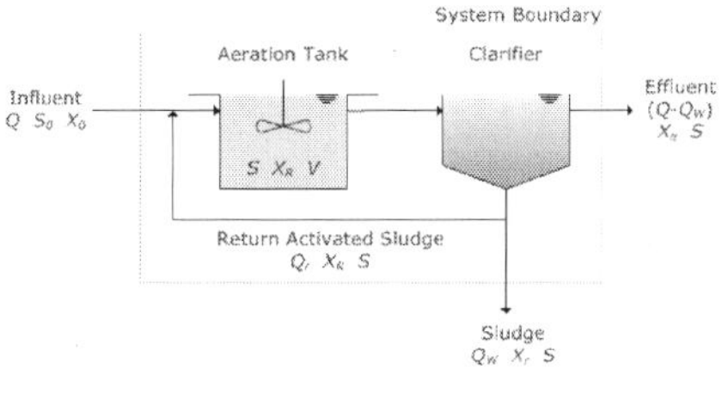

Figure 4.2: The design of the Activated sludge process.

[Source: https://www.lenntech.com/wwtp/wwtp-activated-sludge-process.htm]

Activated sludge in an aeration tank is sludge particles concentrated in wastewater by the development of organisms. The primary sludge consists of several living organisms that sustain themselves on the incoming wastewater whereas; the word 'activated' in activated sludge comes from the fact that it is overflowing with only protozoa, bacteria, and fungi (Geenens et al., 2001). Mixed liquor is a by-product of the combination of differing quantities of recycled liquid that consists of a high ratio of organisms obtained from the screened secondary clarifying tank with screened wastewater. To keep the solids in the form of a suspension, it is supplied with air by stirring and injecting the blend with a significant amount of air. Sometime later, the mixed liquor moves to the clarifier, where it settles down. Partially refined water is passed on for additional treatment as a portion of bacteria is cleared as the liquor settles. Consequently, the settled solids, the *activated sludge,* are moved back to the first tank to start the process again. Numerous alternatives to the basic process are present today. Detail of the three standard variations is included in the issue of *Pipeline:* oxidation ditches extended aeration and sequencing batch reactors. For larger installations activated sludge plant is a highly approved biological treatment process. The terms used in the activated sludge process have been explained in the glossary below (Fuhs& Chen, 1975):

Aerobic – a state when oxygen is present

BOD – biological oxygen demand. Amount of oxygen, needed by the organic matter in the water.

Bulking – Filamentous bacteria does not let the sludge settle down completely, thereby causing the sludge to produce clouds in the secondary clarifiers.

F: M – Food to microbe ratio

Floc – clumps of bacteria

Flocculation – shaking wastewater to produce the minute, suspended particles to clump together into heavier particles (floc) and settle out.

Loading – an amount of material added to the procedure at one time

MLSS – mixed-liquor suspended solids

MLVSS – volatile mixed-liquor suspended solids

Mixed liquor – raw wastewater blended with activated sludge

Package plant – pre-manufactured treatment facility small areas or individual properties utilized to refine wastewater SRT – solids retention time

Sludge – the solids that settle out during the process

Supernatant – The liquid that is suspended from settled sludge. Generally, it concerned with liquid between the sludge present on the base and the scum on the outside.

TSS – total suspended solids

Wasting – Clearing surplus microorganism's small package plants being utilized presently. These plants can generate superior quality effluent for the price. Comparatively little requirement of land and minimal construction costs are some other benefits of the activated sludge process. This process is increasingly being utilized by big cities and societies where an abundant amount of wastewater needs to be highly refined economically. For isolated facilities like small districts, hotels, subunits, hospitals, etc activated sludge process plants are a good option (Dignac et al., 2000; Abou-Elela et al., 2010).

4.2 BIOLOGICAL PROCEDURES

A basic activated sludge process includesnumerous interconnected components(Henze et al., 1999; Zeng et al., 2016):

1. An aeration tank where the biological reactions take place.
2. An aeration source that delivers oxygen and blending.
3. Clarifier, a tank, where solids settle down and are removed from treated wastewater.
4. A method used to either eliminate the solids from the process (waste activated sludge [WAS]) or assemble them to send them back to the aeration tank (return activated sludge [RAS]). Due to the presence of an adequate amount of oxygen and food, the bacteria multiply quickly and grow as they continue moving through the aeration tank(Yasui& Shibata, 1994; Zupanc et al., 2013). The bacteria use almost all the organic material to create new cells by the time the waste makes it to the end of the tank (which takes about to eight hours). After the organisms have settled at the base of the tank, the sludge is added to the tank again to be merged with the incoming wastewater or it is eliminated from the procedure as excess, a practice known as wasting. The supernatant, a fairly clear liquid on top of the sludge, is moved for further treatment as per the requirement (Gujer et al., 1995; Yavuz&Celebi, 2000).

Chemical oxidation can be used to measure organic carbon. By heating the sample with strong sulphuric acid which also contains potassium dichromate, causes the carbon to oxidize, depending on the specified quantity of dichromate utilized in the reaction. The process is called Chemical Oxygen Demand (COD), the by-product is expressed in units of oxygen instead of carbon. Even though it happens to be an analytically simple procedure, its value contains the amount of organic carbon combination that cannot be oxidized biologically which happens to be its drawback. Subsequently, certain aromatic combinations, for instance, a few pyridine, toluene and benzene that can be degraded by bacteria are partially oxidized in the COD process. Generally, the COD can overestimate the carbon that can be eliminated by the activated sludge (Garrett Jr, 1998; Adav et al., 2007).

The 5-day Biological Oxygen Demand (BOD5) is presently used to describe the biodegradable carbon. It is the degree of oxygen consumed by a tiny "seed" of bacteria when restricted, in the dark, inside a bottle consisting of wastewater. Meanwhile,biodegradable organic carbon is obtained and a little carbon is utilized for the aspiration of bacteria causing 4 respective reductions to take place in the dissolved oxygen. However, in the COD test, the biodegradable carbon is portrayed in oxygen units(Artan et al., 1995; Orhon, 1995). It's essentially used for regulating the composition of waste from the treatment water. Since the test was designed to measure the evacuated oxygen in the receiving waters that are affected by the residual decomposable carbon in the sewage. BOD5 is of greatimportance in addition to understanding the organic loading of the influents for process management as the 5 days are crucial to the measurement. Through a brief test (BODST) that can last over half an hour to several hours, steps are being taken to modifying the use of BOD5 as an estimator of influent strength(Luthy et al., 1980).

Two reasons behind the decrease in values obtained from BOD5 than those from COD are listed below:

1. COD test is capable of chemically oxidizing certain combinations that the bacteria in the activated sludge are incapable of degrading.
2. BOD is only calculating the biodegradable carbon that the bacteria oxidize as few carbons removed during the test are not oxidized.

The proportion of BOD5/COD will be related to the constitution of wastewater. The ratio of about 0.5 - 0.6 is seen in wastewaters received from distillery, slaughterhouse, rubber and dairy industries and household sewage, while the ratio for the effluent retiring from the treatment plant is about 0.2. The reason being the- 'hard' BOD, during the treatment the easily biodegradable organic carbon is removed, departing behindthe elements that are not easilydegraded by the bacteria. These will not be quickly degraded or removed by the bacteria in the BOD bottle however they will be easily quantified through chemical oxidation (Joss et al., 2006).

4.3 NATURE OF SLUDGE

Plant operators are possible to navigate the efficiency of the treatment plant's process by evaluating the various attributes of the activated sludge or the sludge quality. Through the regular analysis of the functional and laboratory data; damage control, keeping well maintained and correct records and solving problems before they get serious can assist in carrying out effective operations. A variety of physical, laboratory and visual tests are suggested, which mainly include, evaluation of different chemical conditions, microscopic recognition of prominentkinds of bacteria and a jar test for settle ability and floc performance. The treatment environment .i.e. pH, dissolved oxygen, sludge age, food alterations, total dissolved solids, and temperature, directly affect microbes (Trgovcich et al., 1983; Watt et al., 1985).

A wastewater treatment's activated sludge from the aeration laver works in a complicated network of competing organisms. With about 300 prevalent species, bacteria are the most dominant organisms. They are the most numerous and smallest living things with a cell size ranging from 0.5 – 2 μm. Bacteria have an outer membrane that controls the passage of molecules and ions coming in and out of the cell from the surrounding liquid. A sugar polymer produces a strong cell wall that encompasses the cell membrane. Inside the cell is present the cytoplasm and hundreds of different chemicals whose reactions are controlled by the activity of enzymes. These cells do not possess a nucleus. They are present in different shapes such as spiral, rod-shaped however most of them are spherical. Filamentous bacteria include long chains of minuscule bacterial cells which are occasionally enveloped by a tubular sheath, capable of reaching 100μm length.

Through the cell wall, compounds that have a small molecular weight scatter in the bacteria while bigger and complex molecules that the bacteria compose move out, this is known as secretion. These secretions consist of gels and slime that keep the bacteria and its enzymes together. Enzymes make small monomers that are appropriate for ingestion by disintegrating the larger organic molecules(Aboobakar et al., 2013).

During the growth process, the ingested molecules are utilized by bacteria to produce newer ones. After achieving their standard size, a bacterium splits into two and then repeats the process. In the presence of abundant nutrient molecules, the bacteria start to grow at a quicker pace.

Heterotrophic and autographic bacteria are both found in the wastewater treatment plant, with carbonaceous or heterotrophic being the prominent ones in the group of organisms. Their classification depends upon their nutritional behavior, particularly on organic carbon molecules instead of inorganic ones. On the other hand, autographs utilize inorganic chemicals to produce organic compounds. The nitrifying bacteria in this group are considered important as they eliminate ammonia from the wastewater. There are comparatively fewer species of autographic bacteria that grow at a slower pace; hence they are out-numbered by heterotrophs at developing at a quicker rate (Van Limbergen et al., 1998; Adonadaga, 2015).

4.4 BIOLOGICAL FOUNDATIONS OF WASTEWATER TREATMENT

A secondary treatment method of suspended growth, the activated sludge, basically eliminates non-settle-able, organic, and settle-able suspendable solids. The activated sludge consists of a collection of sludge remnants and microorganisms that are naturally present in unprocessed or settled wastewater.

Developed in aeration tanks these organisms are supplied food and soluble oxygen through the waste. The word "activated" signifies the particles that are swarming with protozoa, bacteria, and fungi (Chapman et al., 1976; Palm et al., 1980). Similar to other treatment facilities, an activated sludge treatment plant's initial treatment procedures on the received wastewater focus on the eradication of the rough or heavy inorganic solids (grit) along with other remains like boards and rags. Most of the settleable and floatable organic matter is eliminated by primary clarifiers (if they are available) (Wang et al., 2017; Droste&Gehr, 2018).

When the effluent moves into the aeration basin, the microbes of the activated sludgeutilize the solids in the wastewater. Following theaeration basin, the microbes and the solids of the wastewater are removed from the water by the mechanism of gravity settling taking place in the secondary clarifier. The sediment and microorganisms are drained back to the anterior of the aeration basin as the clarified water is moved to the next section

(McGhee & Steel, 1991; Kerri, 2003). The design of the activated sludge system has been depicted in Fig. 4.1 (Agridiotis et al., 2007).

4.4.1 Allocating Manageable Influent Feeding

All the functioning aeration basins of the activated sludge systems need to be evenly loaded when the wastewater is fed in them. The front of the aeration basins should be provided with well-made flow splitter boxes that should be routinely checked to make sure that the arrangement of the flow is split according to the required(Feng et al., 2009; Liu et al., 2011).Supplying wastewater throughout different sites of the aeration tank is suitable in most situations, which is called step feeding. This procedure helps in reducing the high requirement of oxygen that may occur where the RAS and influent flow are enclosed in the aeration basin. A disadvantage of this process is that few of the soluble solids in the affluent might pass very quickly through the aeration basin and end up in the wastewater as BOD (Parkin& McCarty, 1981; Yuan et al., 2011).

4.4.2 RegulatingSuitable Dissolved Oxygen and Mixing Levels

To obtain energy for growth and development the microbes in the activated sludge require oxygen to oxidize waste. Lack of oxygen can cause facultative organisms to perform ineffectively, kill or retard the development of aerobic organisms eventually leading to the creation of foul-smelling results of anaerobic analysis. The gradual increase in the number of organisms in the aeration tanks leads to an increase in the quantity of oxygen required to sustain them (Neethling et al., 1985; Hwang & Tanaka, 1998). A high affluent flow or high levels of BOD in the affluent can enhance the organism's activity, therefore, raising the need for oxygen. Hence an adequate amount of oxygen should always be supplied in the aeration tank to guarantee proper waste stabilization. Thus, the amount of oxygenin the aeration tank is also a critical control provided to the operator (Pitt & Jenkins, 1990; Pagilla et al., 1996). For primary activated sludge operations the least dissolved oxygen (D.O.) level of 1.0 mg/L is recommended. Retaining > 1.0 mg/L of D.O. provides a desirable environment for the organisms, resulting in appropriate kinds of organisms and acceptable levels of activity (Daigger et al., 1985; Steffens &Lant, 1999). Unwanted organisms like filamentous type bacteria may develop and overcome the process if the D.O. in the aeration tank is permitted to fall too low for an extended amount of time. Meanwhile, issues like flock fragments being floated on the top of the secondary clarifiers mat

Description	Tapered aeration is mostly attained by using various distances of air diffusers throughout the length of the tank. It is an alteration of the conventional plug-flow system. Based on the requirement of oxy-gen, different aeration rates are used along the length of the tank. When the mixed liquor reaches the efflu-ent end, an increased quantity of air is inserted to the aeration tank's head end.	It is a modified blend of high volumetric loadings with high mixed liquor suspended solids (MLSS) concentrations. Proper mixing is extremely essential. The mixture allows reduced mean cell-resi-dence times and elevated F/M ratios with decreased hydraulic detention times.
BOD removal efficiency (Percentage)		75-90
Process or process Modification	Modified aeration	Kraus process
Description	Except for its brief aeration time and utilization of greater F/M ratios, it is much like the conventional plug-flow process. BOD removal efficiency is lesser as compared to other activated-sludge processes.	Processing waste-water with decreased levels of nitrogen, this process is a modification of the step aeration process. Digester supernatant is provided as a source of nutrition to a segment of the return sludge in a different aeration tank structured to nitrify. The product, mixed liquor, is then inserted into the main plug-flow aeration set-up.
BOD removal efficiency (Percentage)	60-75	85-95
Process or process Modification	Step-feed aeration	Complete mix
Description	Step feed is a variation of the con-ventional plug-flow method which involves the insertion of settled waste-water at numerous sites in the aeration tank to balance the food to micro-organism (F/M) ratio, hence declining the peak oxygen require-ment Generally; three or more equidistant channels are used. The flexibility of execution is one of the major characteristics of this process.	The implementation of the flow plan of a continuous-flow stirred tank reactor. Recovered activated sludge and settled waste-water are usually added at various locations in the aeration tank. All across the tank, the oxygen demand and organic load remain the same.

BOD removal efficiency (Percentage)	85-95	85-95
Process or process Modification	Contact stabilization	High-purity oxygen
Description		In place of air, this process uses high-purity oxygen in the activated-sludge process. The oxygen is released into enclosed aeration tanks to be circulated again. To decrease the amount of carbon dioxide a fragment of gas is expended. pH regulation may be needed. The quantity of oxy-gen supplied is about four times more than the quantity that can be supplied by conventional aeration procedures.
BOD removal efficiency (Percentage)	80-90	85-95
Process or process Modification	Separate-stage Nitrification	Oxidation ditch
Description	For nitrification, this process requires a separate reactor that functions on a feed waste from the previous biological treatment sec-tion. The benefit of this process is that it can be adjusted to satisfy the nutritional requirements.	Mechanical aeration devices are fit in the oxidation ditch which has oval-shaped channels or rings. After the ditch is filled with screened waste-water, it is aer-ated, and circulated for approxi-mately 0.8 to 1.2 ft/s (0.25 to 0.35 m/s). Usually, oxidation ditches function in an extended aeration mode characterized by lengthy detention and solids retention periods. Mostly, secondary sedi-mentation tanks are utilized for this operation.
BOD removal efficiency (Percentage)	85-95	75-95

Process or process Modification	Single-stage Nitrification	Sequencing batch reactor
Description	The decline in ammonia and BOD takes place in a single biological phase. A sequence of either complete-mix reactors or plug-flow can be utilized as reactor configurations.	This reactor is a fill-and-draw kind of reactor system that includes a single complete-mix reactor in which every step of the activated-sludge process takes place. Mixed liquor is present in the reactor throughout all cycles exterminating the need for separate secondary sedimentation tanks.
BOD removal efficiency (Percentage)	85-95	85-95
Process or process Modification	Deep shaft reactor	
Description	The deep vertical shaft reactor a type of the activated-sludge process. It consists of a vertical shaft nearly 400 to 500 ft (120 to 150 m) deep which is used in place of primary clarifiers and aeration basin. The shaft is covered with a steel shell and equipped with a concentric pipe to make an annular reactor. Through the mid of the shaft, air and mixed liquor rushed down and permitted to rise above through the annulus.	
BOD removal efficiency (Percentage)	85-95	

REFERENCES

1. Aboobakar, A., Cartmell, E., Stephenson, T., Jones, M., Vale, P., & Dotro, G. (2013). Nitrous oxide emissions and dissolved oxygen profiling in a full-scale nitrifying activated sludge treatment plant. *Water research*, *47*(2), 524-534.
2. Abou-Elela, S. I., Kamel, M. M., & Fawzy, M. E. (2010). Biological treatment of saline wastewater using a salt-tolerant microorganism. *Desalination*, *250*(1), 1-5.
3. Adav, S. S., Chen, M. Y., Lee, D. J., & Ren, N. Q. (2007). Degradation of phenol by aerobic granules and isolated yeast Candida tropicalis. *Biotechnology and bioengineering*, *96*(5), 844-852.
4. Adham, S. S., Jacangelo, J. G., & Laîné, J. M. (1996). Characteristics and costs of MF and UF plants. *Journal-American Water Works Association*, *88*(5), 22-31.
5. Adonadaga, M. G. (2015). Effect of dissolved oxygen concentration on morphology and settleability of activated sludge flocs. *Journal of Applied & Environmental Microbiology*, *3*(2), 31-37.
6. Agridiotis, V., Forster, C. F., & Carliell-Marquet, C. (2007). Addition of Al and Fe salts during treatment of paper mill effluents to improve activated sludge settlement characteristics. *Bioresource Technology*, *98*(15), 2926-2934.
7. Ahansazan, B., Afrashteh, H., Ahansazan, N., & Ahansazan, Z. (2014). Activated sludge process overview. *International Journal of Environmental Science and Development*, *5*(1), 81.
8. Al-Mutairi, N. Z. (2009). Aerobic selectors in slaughterhouse activated sludge systems: A preliminary investigation. *Bioresource technology*, *100*(1), 50-58.
9. Altalhi, T., Kumeria, T., Santos, A., & Losic, D. (2013). Synthesis of well-organised carbon nanotube membranes from non-degradable plastic bags with tuneable molecular transport: Towards nanotechnological recycling. *Carbon*, *63*, 423-433.
10. American Public Health Association, American Water Works Association, Water Pollution Control Federation, & Water Environment Federation. (1920). *Standard methods for the examination of water and wastewater (Vol.1, pp. 1-33)*. American Public Health Association.
11. American Water Works Association (Ed.). (1990). *Water quality and treatment: A handbook of community water supplies (Vol.1, pp. 1-33)*.

McGraw-Hill Companies.

12. Arévalo, J., Moreno, B., Pérez, J., & Gómez, M. A. (2009). Applicability of the Sludge Biotic Index (SBI) for MBR activated sludge control. *Journal of hazardous materials*, *167*(1-3), 784-789.
13. Artan, N., Orhon, D., & Sözen, S. (1995). Conceptual basis of wastewater nitrogen removal in sensitive coastal areas. *Water Science and Technology*, *32*(11), 77-84.
14. Balkema, A. J., Weijers, S. R., & Lambert, A. J. D. (1998). Sustainability criteria for the comparison of wastewater treatment technologies. In *The 11th European junior scientist workshop (Vol.17, pp. 1-53), February 12-15, 1998, Wildpark Eekholt, Germany*. Fachhochschule Lübeck.
15. Barnard, J. L. (1976). A review of biological phosphorus removal in the activated sludge process. *Water Sa*, *2*(3), 136-144.
16. Barnard, J. L. (1976). A review of biological phosphorus removal in the activated sludge process. *Water Sa*, *2*(3), 136-144.
17. Bataineh, F., Najjar, M., & Malkawi, S. (2002). Wastewater Reuse, in water demand management forum. *Amman (Vol.1, pp. 1-40), Jordan.*
18. C. Reid, J. M., & Nason, R. B. (1993). Automatic monitoring of activated sludge settling characteristics and plant control applications. *Water and Environment Journal*, *7*(6), 636-645.
19. Chapman, T. D., Matsch, L. C., & Zander, E. H. (1976). Effect of high dissolved oxygen concentration in activated sludge systems. *Journal (Water Pollution Control Federation)*, 1, 2486-2510.
20. Chiesa, S. C., Irvine, R. L., & Manning Jr, J. F. (1985). Feast/famine growth environments and activated sludge population selection. *Biotechnology and bioengineering*, *27*(5), 562-568.
21. Cole, C. A., Stamberg, J. B., & Bishop, D. F. (1973). Hydrogen peroxide cures filamentous growth in activated sludge. *Journal (Water Pollution Control Federation)*, 1, 829-836.
22. Collivignarelli, C., & Bertanza, G. (1999). Simultaneous nitrification-denitrification processes in activated sludge plants: Performance and applicability. *Water Science and Technology*, *40*(4-5), 187-194.
23. Copeland, C. (1999). Clean Water Act: a summary of the law. Washington, DC: Congressional Research Service, Library of Congress, 1, 1-100.
24. Daigger, G. T., Robbins Jr, M. H., & Marshall, B. R. (1985). The design

of a selector to control low-F/M filamentous bulking. *Journal (Water Pollution Control Federation)*, 1, 220-226.

25. de Oliveira, L. L., Duarte, I. C. S., Sakamoto, I. K., & Varesche, M. B. A. (2009). Influence of support material on the immobilization of biomass for the degradation of linear alkylbenzene sulfonate in anaerobic reactors. *Journal of Environmental Management, 90*(2), 1261-1268.
26. Debusscher, D., Vanhooren, H., & Vanrolleghem, P. A. (1999). Benchmarkingtwobiomassloadingcontrolstrategiesforactivatedsludge WWTPs. *MEDEDELINGEN-FACULTEIT LANDBOUWKUNDIGE EN TOEGEPASTE BIOLOGISCHE WETENSCHAPPEN, 64*, 127-132.
27. Dignac, M. F., Ginestet, P., Rybacki, D., Bruchet, A., Urbain, V., & Scribe, P. (2000). Fate of wastewater organic pollution during activated sludge treatment: nature of residual organic matter. *Water Research, 34*(17), 4185-4194.
28. Directive, E. U. W. (1991). Council Directive of 21. May 1991 concerning urban waste water treatment (91/271/EEC). *J. Eur. Commun, 34*, 40.
29. Directive, W. F. (2003). Common implementation strategy for the water framework directive (2000/60/EC). *Guidance document*, (7), 1-33.
30. Droste, R. L., & Gehr, R. L. (2018). *Theory and practice of water and wastewater treatment (Vol.1,pp. 1-22)*. John Wiley & Sons.
31. Faeth, P. (2000). *Fertile ground: nitrient trading's potential to cost-effectively improve water quality* (No. 363.73946 F149, pp. 1-33). World Resources Institute, Washington, DC (EUA).
32. Feng, X., Lei, H., Deng, J., Yu, Q., & Li, H. (2009). Physical and chemical characteristics of waste activated sludge treated ultrasonically. *Chemical Engineering and Processing: Process Intensification, 48*(1), 187-194.
33. Franklin, L. B., & Stensen, H. D. (2003). Wastewater Engineering Treatment and Reuse, 1, 1-45.
34. Fuhs, G. W., & Chen, M. (1975). Microbiological basis of phosphate removal in the activated sludge process for the treatment of wastewater. *Microbial Ecology, 2*(2), 119-138.
35. Gaius-obaseki, T. (2010). Hydropower opportunities in the water industry. *International Journal of Environmental Sciences, 1*(3), 392-402.

36. Garrett Jr, M. T. (1998). Instrumentation, control and automation progress in the United States in the last 24 years. *Water science and technology*, *37*(12), 21-25.
37. Geenens, D., Bixio, B., & Thoeye, C. (2001). Combined ozone-activated sludge treatment of landfill leachate. *Water Science and Technology*, *44*(2-3), 359-365.
38. Ghangrekar, M. M., Asolekar, S. R., & Joshi, S. G. (2005). Characteristics of sludge developed under different loading conditions during UASB reactor start-up and granulation. *Water research*, *39*(6), 1123-1133.
39. Gros, M., Petrović, M., Ginebreda, A., & Barceló, D. (2010). Removal of pharmaceuticals during wastewater treatment and environmental risk assessment using hazard indexes. *Environment international*, *36*(1), 15-26.
40. Grünheid, S., Amy, G., & Jekel, M. (2005). Removal of bulk dissolved organic carbon (DOC) and trace organic compounds by bank filtration and artificial recharge. *Water research*, *39*(14), 3219-3228.
41. Gujer, W., Henze, M., Mino, T., Matsuo, T., Wentzel, M. C., & Marais, G. V. R. (1995). The activated sludge model No. 2: biological phosphorus removal. *Water Science and Technology*, *31*(2), 1-11.
42. Hamed, M. M., Khalafallah, M. G., & Hassanien, E. A. (2004). Prediction of wastewater treatment plant performance using artificial neural networks. *Environmental Modelling & Software*, *19*(10), 919-928.
43. Henze, M., Gujer, W., Mino, T., Matsuo, T., Wentzel, M. C., Marais, G. V. R., & Van Loosdrecht, M. C. (1999). Activated sludge model no. 2d, ASM2d. *Water Science and Technology*, *39*(1), 165-182.
44. Hernández-Esparza, M., Doria-Serrano, M. C., Acero-Salinas, G., & Ruiz-Treviño, F. A. (2006). Removal of high phenol concentrations with adapted activated sludge in suspended form and entrapped in calcium alginate/cross-linked poly (N-vinyl pyrrolidone) hydrogels. *Biotechnology progress*, *22*(6), 1552-1559.
45. Hwang, Y., & Tanaka, T. (1998). Control of Microthrix parvicella foaming in activated sludge. *Water Research*, *32*(5), 1678-1686.
46. Joss, A., Zabczynski, S., Göbel, A., Hoffmann, B., Löffler, D., McArdell, C. S., ... & Siegrist, H. (2006). Biological degradation of pharmaceuticals in municipal wastewater treatment: proposing a classification scheme. *Water research*, *40*(8), 1686-1696.

47. Juang, D. F., & Morgan, J. M. (2001). The applicability of the API 20E and API Rapid NFT systems for the identification of bacteria from activated sludge. *Electronic Journal of Biotechnology*, *4*(1), 1-2.

48. Kerri, K. D. (2003). *Operation of wastewater treatment plants: A field study training program* (Vol. 2, pp. 1-44). California State University.

49. Kianmehr, P., Parker, W., & Seto, P. (2013). Assessment of waste-activated-sludge pretreatment by ultrasound and applicability of biodegradability indicators. *Journal of Environmental Engineering*, *139*(4), 545-553.

50. Lee, M. G., Kam, S. K., & Suh, K. H. (2012). Adsorption of non-degradable eosin Y by activated carbon. *Journal of Environmental Science International*, *21*(5), 623-631.

51. Lee, W., Kang, S., & Shin, H. (2003). Sludge characteristics and their contribution to microfiltration in submerged membrane bioreactors. *Journal of membrane science*, *216*(1-2), 217-227.

52. Lesouef, A., Payraudeau, M., Rogalla, F., & Kleiber, B. (1992). Optimizing nitrogen removal reactor configurations by on-site calibration of the IAWPRC activated sludge model. *Water Science and Technology*, *25*(6), 105-123.

53. Lim, S. J., & Kim, T. H. (2014). Applicability and trends of anaerobic granular sludge treatment processes. *Biomass and bioenergy*, *60*, 189-202.

54. Liu, J. J., Wang, X. C., & Fan, B. (2011). Characteristics of PAHs adsorption on inorganic particles and activated sludge in domestic wastewater treatment. *Bioresource technology*, *102*(9), 5305-5311.

55. Liu, Y. (2003). Chemically reduced excess sludge production in the activated sludge process. *Chemosphere*, *50*(1), 1-7.

56. Liu, Y., & Tay, J. H. (2001). Strategy for minimization of excess sludge production from the activated sludge process. *Biotechnology advances*, *19*(2), 97-107.

57. Luthy, R. G., Tallon, J. T., & Sekel, D. J. (1980). Biological treatment of synthetic fuel wastewater. *Journal of the Environmental Engineering Division*, *106*(3), 609-629.

58. Marrot, B., Barrios-Martinez, A., Moulin, P., & Roche, N. (2006). Biodegradation of high phenol concentration by activated sludge in an immersed membrane bioreactor. *Biochemical Engineering Journal*, *30*(2), 174-183.

59. McGhee, T. J., & Steel, E. W. (1991). *Water supply and sewerage* (Vol. 6, pp. 1-50). New York: McGraw-Hill.
60. Metcalf, L., Eddy, H. P., & Tchobanoglous, G. (1979). *Wastewater engineering: treatment, disposal, and reuse (Vol.1, pp. 1-40)*. New York: McGraw-Hill.
61. Mohan, S. V. (2009). Harnessing of biohydrogen from wastewater treatment using mixed fermentative consortia: process evaluation towards optimization. *International Journal of Hydrogen Energy*, *34*(17), 7460-7474.
62. Murakami, T., Usui, J., Takamura, K., & Yoshikawa, T. (2000). Application of immersed-type membrane separation activated sludge process to municipal wastewater treatment. *Water science and technology*, *41*(10-11), 295-301.
63. Nagaoka, H., Ueda, S., & Miya, A. (1996). Influence of bacterial extracellular polymers on the membrane separation activated sludge process. *Water Science and Technology*, *34*(9), 165-172.
64. Neethling, J. B., Jenkins, D., & Johnson, K. M. (1985). Chemistry, microbiology, and modeling of chlorination for activated sludge bulking control. *Journal (Water Pollution Control Federation)*, 1, 882-889.
65. Oh, J., & Silverstein, J. (1999). Oxygen inhibition of activated sludge denitrification. *Water Research*, *33*(8), 1925-1937.
66. Orhon, D. (1995). Scientific basis for wastewater treatment and disposal in Istanbul. *Water Science and Technology*, *32*(2), 199.
67. Pagilla, K. R., Jenkins, D., & Kido, W. H. (1996). Nocardia control in activated sludge by classifying selectors. *Water environment research*, *68*(2), 235-239.
68. Palm, J. C., Jenkins, D., & Parker, D. S. (1980). Relationship between organic loading, dissolved oxygen concentration and sludge settleability in the completely-mixed activated sludge process. *Journal (Water Pollution Control Federation)*, 1, 484-2506.
69. Palm, J. C., Jenkins, D., & Parker, D. S. (1980). Relationship between organic loading, dissolved oxygen concentration and sludge settleability in the completely-mixed activated sludge process. *Journal (Water Pollution Control Federation)*, 1, 2484-2506.
70. Parkin, G. F., & McCarty, P. L. (1981). A comparison of the characteristics of soluble organic nitrogen in untreated and activated

sludge treated wastewaters. *Water Research, 15*(1), 139-149.

71. Pitt, P., & Jenkins, D. (1990). Causes and control of Nocardia in activated sludge. *Research Journal of the Water Pollution Control Federation*, 1, 143-150.
72. Pitter, P. (1976). Determination of biological degradability of organic substances. *Water research, 10*(3), 231-235.
73. Ramanathan,M.,&GaudyJr,A.F.(1971).Steady-statemodelforactivated sludge with constant recycle sludge concentration. *Biotechnology and Bioengineering, 13*(1), 125-145.
74. Sakai, Y., Fukase, T., Yasui, H., & Shibata, M. (1997). An activated sludge process without excess sludge production. *Water Science and Technology, 36*(11), 163.
75. Shaul, G. M., Holdsworth, T. J., Dempsey, C. R., & Dostal, K. A. (1991). Fate of water soluble azo dyes in the activated sludge process. *Chemosphere, 22*(1-2), 107-119.
76. Smulders, E., & Rähse, W. (2002). *Laundry detergents* (Vol.1, p. 52). Weinheim: Wiley-Vch.
77. Spanjers, H., & Vanrolleghem, P. (1995). Respirometry as a tool for rapid characterization of wastewater and activated sludge. *Water Science and Technology, 31*(2), 105-114.
78. Sridang, P. C., Wisniewski, C., Ognier, S., & Grasmick, A. (2006). The role of the nature and composition of solutions/suspensions in fouling of plane organic membranes in frontal filtration: Application to water and wastewater clarification. *Desalination, 191*(1-3), 71-78.
79. Steffens, M. A., & Lant, P. A. (1999). Multivariable control of nutrient-removing activated sludge systems. *Water Research, 33*(12), 2864-2878.
80. Sustarsic, M. (2009, November). Wastewater treatment: Understanding the activated sludge process. In *Safety in ammonia Plants and related facilities symposium* (Vol.1, pp. 26-29).
81. Trgovcich, B., Kirsch, E. J., & Grady Jr, C. L. (1983). Characteristics of activated sludge effluents before and after breakpoint chlorination. *Journal (Water Pollution Control Federation)*, 966-976.
82. Van Limbergen, H., Top, E. M., & Verstraete, W. (1998). Bioaugmentation in activated sludge: current features and future perspectives. *Applied Microbiology and Biotechnology, 50*(1), 16-23.
83. Wang, D., Wang, Y., Liu, Y., Ngo, H. H., Lian, Y., Zhao, J., ... & Li,

X. (2017). Is denitrifying anaerobic methane oxidation-centered technologies a solution for the sustainable operation of wastewater treatment plants?. *Bioresource Technology*, *234*, 456-465.

84. Watt, R. D., Kirsch, E. J., & Grady Jr, C. L. (1985). Characteristics of activated sludge effluents before and after ozonation. *Journal (Water Pollution Control Federation)*, 157-166.
85. Yasui, H., & Shibata, M. (1994). An innovative approach to reduce excess sludge production in the activated sludge process. *Water science and technology*, *30*(9), 11.
86. Yavuz, H., & Celebi, S. S. (2000). Effects of magnetic field on activity of activated sludge in wastewater treatment. *Enzyme and Microbial Technology*, *26*(1), 22-27.
87. Yi, S., Zhuang, W. Q., Wu, B., Tay, S. T. L., & Tay, J. H. (2006). Biodegradation of p-nitrophenol by aerobic granules in a sequencing batch reactor. *Environmental science & technology*, *40*(7), 2396-2401.
88. Yuan, H., Zhu, N., & Song, F. (2011). Dewaterability characteristics of sludge conditioned with surfactants pretreatment by electrolysis. *Bioresource technology*, *102*(3), 2308-2315.
89. Zeng, J., Gao, J. M., Chen, Y. P., Yan, P., Dong, Y., Shen, Y., ... & Zhang, P. (2016). Composition and aggregation of extracellular polymeric substances (EPS) in hyperhaline and municipal wastewater treatment plants. *Scientific reports*, *6*, 26721.
90. Zupanc, M., Kosjek, T., Petkovšek, M., Dular, M., Kompare, B., Širok, B., ... & Heath, E. (2013). Removal of pharmaceuticals from wastewater by biological processes, hydrodynamic cavitation and UV treatment. *Ultrasonics sonochemistry*, *20*(4), 1104-1112.

CHAPTER 5

ANAEROBIC WASTEWATER TREATMENT PROCESS

CONTENTS

color reduction of 75 % and a COD reduction of 79 % was attainable only at the first anaerobic stage that contains a fixed bed filter, operating at 7.7 h HRT (Hirano et al., 2007).

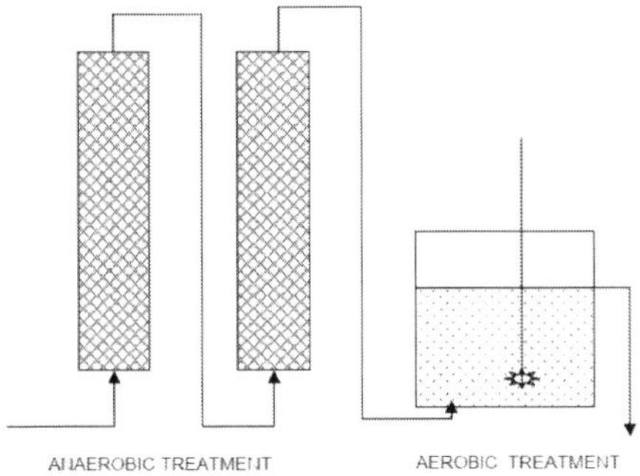

Figure 5.3: Diagrammaticportrayal of anaerobic-aerobic textile wastewater treatment.

[Source: https://www.kth.se/polopoly_fs/1.650925.1550157132!/JPSU16p13.pdf]

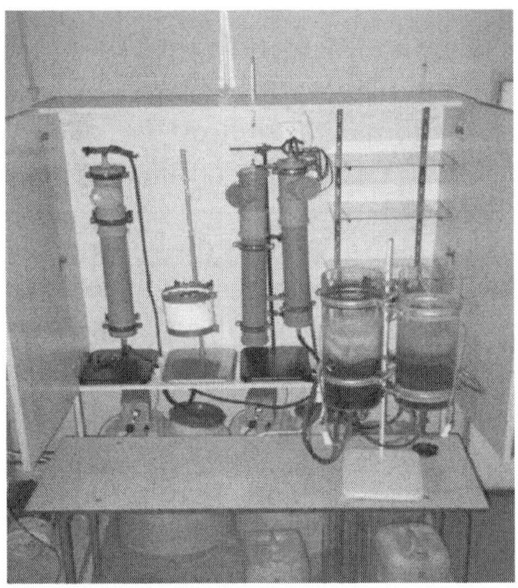

Figure 5.4: Anaerobic-aerobic textile wastewater treatment's laboratory system.

[Source: https://www.kth.se/polopoly_fs/1.650925.1550157132!/JPSU16p13.pdf]

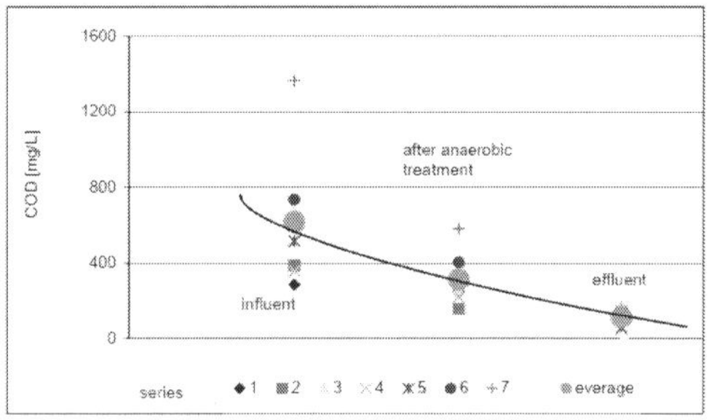

Figure 5.5: Elimination of COD in the anaerobic-aerobic textile wastewater treatment process.

[Source: https://www.kth.se/polopoly_fs/1.650925.1550157132!/JPSU16p13.pdf]

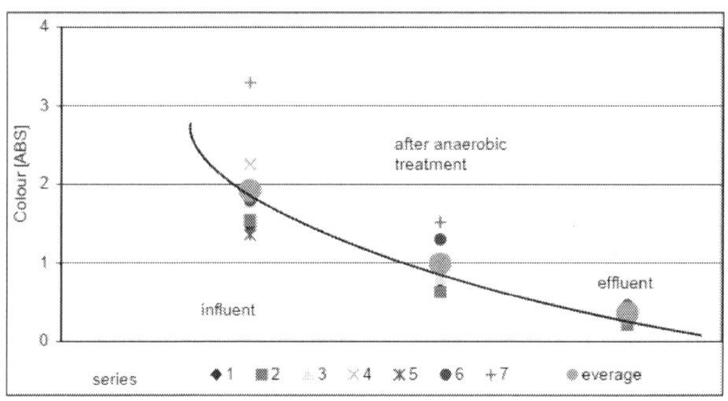

Figure 5.6: eradication of color in the anaerobic-aerobic textile wastewater treatment process.

[Source: https://www.kth.se/polopoly_fs/1.650925.1550157132!/JPSU16p13.pdf]

Kleerebezem et al. (1999) studied the impacts of benzoate and acetate by a syntrophic methanogenic culture on anaerobic degradation of terephthalate (1,4-benzene dicarboxylate). The absolute dependence of the formation of benzoate resulting from decarboxylation of terephthalate on the concomitant fermentation of benzoate was seen by them (Pereira et al., 2005; Bandara et al., 2011).

Van Eekert et al. (1999) proved the significant mechanisms for the transformation of chlorinated ethanes, in a UASB reactor by the methanogenic consortium, to be biotic and co-metabolic conversions.

32. Nykova N., Muller T. G., Gyllenberg M., Timmer J. (2002). Quantitative analyses of anaerobic wastewater treatment processes: identifiability and parameter estimation. *Biotechnology and Bioengineering*, **78**, (1), 89-103.
33. Orozco A. (1997). Pilot and full scale anaerobic treatment of low strength wastewater at sub-optimal temperature (15oC) with a hybrid plug flow reactor.Proceedings of 8th International Conference on Anaerobic Digestion, Sendai, Japan , **2**, 183-191.
34. Pantea E.V., &Romocea T. (2008). Thermophilic anaerobic wastewater treatment. *Protectia Mediului, Analele Universitatii din Oradea*, **8**, 454-458.
35. Parawira W. (2004). Anaerobic treatment of agricultural residues and wastewater. Application of highrate reactors, *Doctoral Dissertation, Department of Biotechnology, Lund University, Sweden.* Razo-Flores, E., Lettinga, G.,& Field, J. A. (1999). Biotransformation and biodegradation of selected nitroaromatics under anaerobic conditions. *Biotechnol. Prog.*, **15**, 3, 358-365.
36. Pereira, L. C. C., Jiménez, J. A., Koening, M. L., Porto Neto, F. F., Medeiros, C., & Costa, R. M. D. (2005). Effect of coastline properties and wastewater on plankton composition and distribution in a stressed environment on the north coast of Olinda-PE (Brazil). *Brazilian Archives of Biology and Technology*, *48*(6), 1013-1026.
37. Petala, M., Samaras, P., Zouboulis, A., Kungolos, A., & Sakellaropoulos, G. (2006). Ecotoxicological properties of wastewater treated using tertiary methods. *Environmental Toxicology: An International Journal*, *21*(4), 417-424.
38. Rebac, S., van Lier, J. B., Lens, P., Stams, A. J., Dekkers, F., Swinkels, K. T. M., & Lettinga, G. (1999). Psychrophilic anaerobic treatment of low strength wastewaters. *Water Science and Technology*, *39*(5), 203-210.
39. Rumana R., Somchai D., Kannitha K. (2000). Anaerobic processes. Literature Review, *Water Environment Research*, **72**, (5), 576-656.
40. Sandars, D. L., Audsley, E., Canete, C., Cumby, T. R., Scotford, I. M., & Williams, A. G. (2003). Environmental benefits of livestock manure management practices and technology by life cycle assessment. *Biosystems Engineering*, *84*(3), 267-281.
41. Singh L., Alam S.I., Ramana K.V. (1999). Effect of fluctuating temperature regime on psychrophilic anaerobic digestion of nightsoil. *Defence Science Journal*, **49**, (2), 135-140.

42. Singh, K. S., & Viraraghavan, T. (1999). Municipal wastewater treatment by UASB process: Start-up at 20° C and operation at mesophilic and psychrophilic temperatures. *Proc. Water Environ. Fed. 72nd Annu. Conf. Exposition (Vol. 72, pp. 1-44)*, New Orleans, LA.
43. Smith, A. L., Skerlos, S. J., & Raskin, L. (2013). Psychrophilic anaerobic membrane bioreactor treatment of domestic wastewater. *Water research*, *47*(4), 1655-1665.
44. Tafdrup, S. (1995). Viable energy production and waste recycling from anaerobic digestion of manure and other biomass materials. *Biomass and Bioenergy*, *9*(1-5), 303-314.
45. Tan N. C., Prenafeta-Boldu F. X., Opsteeg J. L., Lettinga G., Field J. A. (1999). Biodegradation of azo dyes in coculture of anaerobic granular sludge with aerobic aromatics amine degrading enrichment cultures. *Appl. Microbiol.* Biotechnol., **51**, 865-871.
46. Van Eekert M. H., Stams A. J., Field J. A. (1999). Gratuitous dechlorination of chloroethanes by methanogenic granular sludge Gratuitous dechlorination of chloroethanes by methanogenic granular sludge. *Appl. Microbiol. Biotechnol.*, **51**, 46-52.
47. Van Lier, J. B. (2008). High-rate anaerobic wastewater treatment: diversifying from end-of-the-pipe treatment to resource-oriented conversion techniques. *Water Science and Technology*, *57*(8), 1137-1148.
48. Van Lier, J. B., Rebac, S., & Lettinga, G. (1997). High-rate anaerobic wastewater treatment under psychrophilic and thermophilic conditions. *Water science and technology*, *35*(10), 199-206.
49. Vanderloop S. L, Suidan M. T., Moteleb M. A., Maloney S. W. (1999). Biotransformation of 2,4dinitrotoluene under different electron acceptor conditions. *Water Research*, **33**, (5), 1287-1295.
50. Vartak, D. R., Engler, C. R., McFarland, M. J., & Ricke, S. C. (1997). Attached-film media performance in psychrophilic anaerobic treatment of dairy cattle wastewater. *Bioresource Technology*, *62*(3), 79-84.
51. Wilkie, A. C. (2005). Anaerobic digestion: biology and benefits. *Dairy manure management: treatment, handling, and community relations*, 1, 63-72.
52. Yang, K., Yue, Q., Han, W., Kong, J., Gao, B., Zhao, P., & Duan, L. (2015). Effect of novel sludge and coal cinder ceramic media in combined anaerobic–aerobic bio-filter for tetracycline wastewater treatment at low temperature. *Chemical Engineering Journal*, *277*, 130-139.

CHAPTER

6

MEMBRANE BIOREACTOR APPLICATION IN WATER TREATMENT

CONTENTS

6.1 INTRODUCTION

A new kind of wastewater treatment technology is a Membrane bioreactor (MBR). It associates conventional activated sludge (CAS) treatment process with the membrane separation process. MBR can efficaciously provide high-quality clarified water because the size of the used membrane pore may commonly below 0.1 mm.MBR process has acquired additional attention for its advantages like space-saving, less sludge production, and high removal efficiency of pollutants (Melin et al., 2006; Treguer et al., 2010). Microorganisms are stranded into the bioreactor by cause of the function of membrane filtration. Bioreactor gives the chance of effective control of the biological reactions and modifies the situations of the microorganisms in the aerated tank (Choi et al., 2002; Kraume & Drews, 2010). It allows high mixed liquor suspended solids (MLSS), long sludge retention time (SRT) concentration. According to the working mechanism, MBR process, as a whole, can be categorized into three parts: extractive MBR, diffusive MBR, and rejection MBR in which the first two kinds are new MBR processes and are still in the process of developing (Stephenson et al., 2000; Judd, 2010). More research works associated with MBR have been centered on the rejection of MBR for water and wastewater treatment, and less consideration on the other two types of MBR from the past 20 years. In recent days, the MBR process has been brilliantly practiced worldwide, consisting of small-scale industrial WWTPs, drinking water treatment plants, and larger-scale municipal wastewater treatment plants (WWTPs) (Yang et al., 2006; Judd, 2008).

6.1.1 Configurations of Membrane Bioreactor

According to the configuration of analysts, they generally refer to MBR as rejection MBR, which can be categorized into three parts:

1. Immersed/submerged membrane bioreactor (termed to as iMBR). Membrane modules are instantly immersed into the bioreactor, in iMBR (Fig. 6.1). To get the effluent through the membrane, the suction pump is used. During this, sludge is stranded into the bioreactor by the membrane. The operational processes are generally established by the inconsistent types. Air is also used to sustain aerobic circumstances or for yielding oxygen and in turn scour the outer of the membrane and cleaning the membrane surface.Due to lower fouling potentials and minimum energy-intensive, the immersed MBR is more generally utilized than the crossflow MBR (cMBR).

2. The amalgamation of a traditional bioreactor with membrane filtration device into an individual unit process is the cMBR.The secondary settling vessel of a conventional biological treatment system is similar to the membrane module in the MBR, in which liquid and solid are isolated. During this, the slime is transferred into the bioreactor and the permeate is accumulated (Fig. 6.2). In this way, the membrane is well scoured in situ and functioned with high slime concentration in the MBR reactor.
3. Hybrid MBR is equivalent to the iMBR system. In the reactor, it is loaded with some stuffing (Fig. 6.3). Due to the stuffing having the mechanism of normalizing the treatment process productivity and lessening the membrane fouling, this system is considered to be better as compared to the immersed MBR. (Xiao et al., 2014; Bani-Melhem et al., 2015).

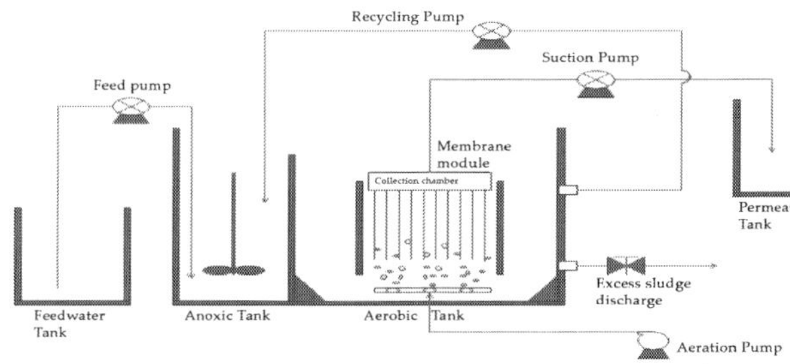

Figure 6.1: Strategical diagram of immersed MBR (iMBR).

[Source:https://www.researchgate.net/figure/Schematic-view-of-the-internal-submerged-MBR_fig1_304746665]

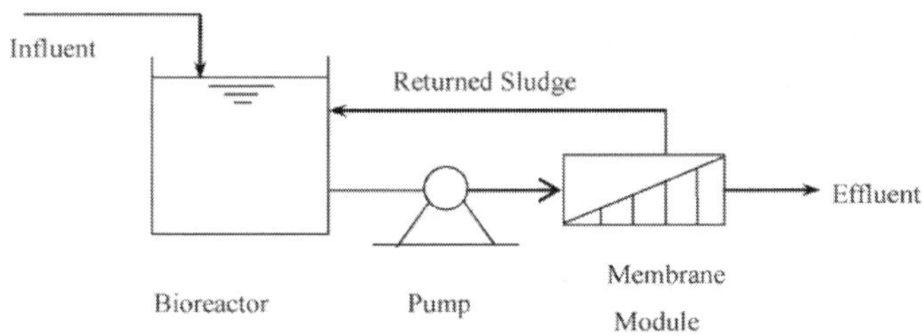

Figure 6.2: Diagrammatic view of crossflow MBR (cMBR).

[Source:https://www.researchgate.net/publication/316141914_Membrane_Bioreactor_Application_in_Water_Treatment]

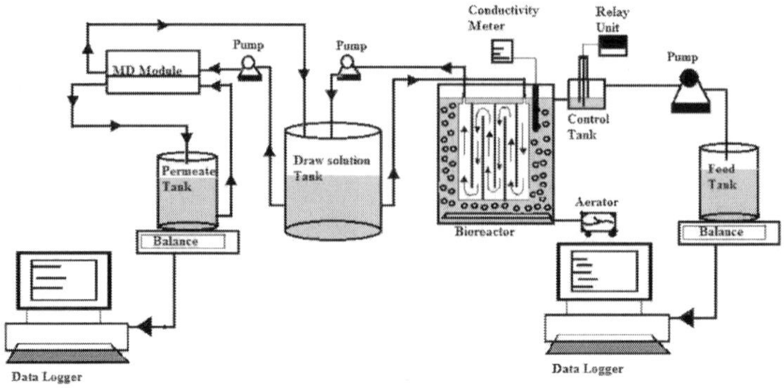

Figure 6.3: Illustrative diagram of hybrid MBR.

[Source: https://www.researchgate.net/figure/Schematic-diagram-of-lab-scale-Hybrid-FO-MBR-MD-system_fig1_332975610]

6.1.2 Characteristics of membrane bioreactor

The MBR process has distinct superiority over conventional wastewater treatment process.2

1. High quality clarified water is produced by MBR. Indicative output condition of MBR systems (ultrafiltration or microfiltration) consists of SSo1 mg/L and contaminant o0.2 NTU (reflecting on the membrane nominal pore size). MBR is the removal of organic matters which is extracted from two factors: one is the membrane filtration of high-molecular weighted organic matters, the other is the biodegradation of organic contaminants in the bioreactor (Fountoulakis et al., 2016).
2. MBR has a tiny footprint. Removal of tertiary sand filtration and secondary settling process is done, thus lessening the plant footprint. In some cases, the footprint can be more lessened since other process units like UV disinfection can also be minimized or removed.
3. MBR process contrasted with the CAS process allows separate SRT and hydraulic retention time (HRT). MBR is harder to operate in the CAS system. In the bioreactor, sludge could be carried by a membrane which can enhance MBR biodegradation productivity and likewise give improved control of HRT and SRT in the system (Chaize & Huyard, 1991)

4. Low excess slime development can be obtained by the MBR which can be constructed with long slime age. It also encourages the nourishment of nitrifying bacteria, thus enhancing nitrogen removal.
5. Since the membrane has an efficacious pore size below 0.1mm—lesser than the viruses and pathogenic bacteria in the slime so MBR renders a hurdle to specific chlorine-resistant pathogens.

6.2 DESIGN AND OPERATION OF MEMBRANE BIO-REACTOR

6.2.1 Design of Membrane Bioreactor

For the design of MBRs, there is not a systematic and deliberate method presently. Choosing the design parameter is generally based on the conclusions of pilot-scale experiments and bench-scale to treat particular wastewater by MBR (Fane et al., 2002; Shim et al., 2002). This portion will precede and compile the design process of MBR.

To configure the MBR is the first basic step; MBR chiefly has two types of commonly utilized configurations, cMBR and submerged.

Submerged MBR is described by a tiny, compact instrument, nonwater cycle, small working pressure, and less energy consumption. During aeration, slime is hard to collect in the membrane surface and plug the membrane pore due to the formulation of turbulence and shear. As a whole, submerged MBR is utilized only for aerobic treatment. The use of aeration mode, the packing density, also the location of the aeration apparatus placed become key factors in the design of MBR systems. To get a higher elimination rate of organic matter, the SRT and HRT of cMBR can be forcefully controlled. Nitrifying bacteria can be nourished due to longer generation time. For anaerobic treatment and aerobic treatment, cMBR can be utilized. (Ivanovic & Leiknes, 2012).

There are three sections to be constructed in MBR design, which will be introduced simultaneously as given(Sofia et al., 2004; Wei et al., 2004): (i) the design of bioreactor; (ii) the selection and design of membrane module; (iii) aeration system design.

1. The design of the bioreactor
   (a) In the bioreactor, determination of the organic matter-slime loading rate (Ns)

The eliminated amounts of organic matter per unit of slime and time are termed as organic matter-slime loading rates (Ns). It is an essential control parameter in the operation of activated slime design and process, which is fundamental to verify the reasonable size of the project and assure the system treatment quality.

Because of the continuation of the membrane modules, bioreactor retaining a very high slime composition which is 5–10 times greater than the CAS process in the MBR operation. The treatment capability in the system is excellent only if a lower organic matter-slime loading rate is selected but there is a corresponding gain in the footprint of the infrastructure investments and treatment works.

Anyhow, the treatment quality may be influenced if we select greater Ns, although lessening the area of infrastructure investment and treatment works. Hence, in the MBR process design, it is a major problem to select reasonable Ns. Commonly, Ns in the reach of 0.3–0.4 kg COD/(kg VSS.d) is appropriate.

(b) Determine the slime concentration in the bioreactor

Generally, we can determine the better suitable amount of activated slime proportion (X) through analysis because of its complication to get through calculated values. Many courses offered the range of X values from 6000 to 20000 mg/L, reflecting on the concentration of organic matter in raw water. X value uses the low value for the analysis of less concentration organic wastewater; in reverse, X value utilizes the high value (Mutamim et al., 2013).

(c) According to the given established Ns, X values, likewise the influent efficiency and effluent standards, compute the volume of the size of the bioreactor

$$V = \frac{Q\left(S_0 - S_e\right)}{XN_s} \qquad (1)$$

Where Se is the effluent concentration (mg/L), S0 is the influent concentration (mg/L), X is slime concentration, V is the volume of the bioreactor (m3) and Q is the flux.

2. Membrane module selection and design

(d) Choose the kind of membrane modules

The membrane is polluted immediately due to the high concentration of wastewater going through it, in MBR. Therefore, it is critical to choose a suitable membrane module and antifouling membrane.

The membrane is categorized into an inorganic membrane and organic membrane which is established on the dissimilar membrane materials (Yu et al., 2006). Ordinary organic membrane materials consist of polyether sulfone (PES), polyacrylonitrile (PAN), polyvinylidene fluoride (PVDF), polypropylene (PP), polyethylene (PE), polysulfone (PS), and so on. The inorganic membrane comprises chiefly ceramic materials, metal oxides, and metals, etc. Few drawbacks of an organic membrane can be controlled by an inorganic membrane, such as high flux, chemical stability, marvelous mechanical properties, and resistance to pollution, easy to scour, etc.

There are two kinds of membrane modules used for MBR, presently. These are tubular membrane and hollow fiber membrane, in which tubular membrane is chiefly used for cMBR and the hollow fiber membrane is chiefly used for submerged MBR. Tubular membrane module has the advantage of hard to plug, comfortable scouring, and convenience for treatment of sewage, low pretreatment demands for liquid with good hydrodynamic conditions but it has the drawback of high expense. Hollow fiber membrane has the advantage of low expense, high packing density, high anti-pressure, and the disadvantage of easily sealed.

(e) Design the needed area of membrane modules

By the size and design membrane module package form and membrane output per module, calculate the area of the membrane module.

(f) Choose and design of the form of flux output

Two kinds of flux output from the membrane module are used: inside-out and outside-in. As a matter of fact, in applied engineering, outside-in MBR is usually utilized, because of the inside-out MBR having the disadvantage of the lesser flow channel and being effortlessly sealed.

3. Aeration system design

The first constituent of aeration involves the bioreactor and especially, the demand of the blended liquor for the air needed for (a) dissolved oxygen (DO) for retaining a workable microorganism population for biotreatment, and (b)solids agitation, however, the first constituent of aeration involves membrane unit, which is important to activate the membrane unit in an MBR to clean solids from the membrane.

By using the mass balance on the system, we can find the demand of the oxygen to sustain a community of microorganisms and ammonia and nitrite to nitrate and degrade BOD (Tchobanoglous et al., 2003):

$$m_0 = Q(S - S_0) - 1.42P_x + 4.33Q(NO_x) - 2.83Q(NO_x) \quad (2)$$

where m0 is the total demanded of oxygen (g/day). The first term in Eq. (2) describes to substrate oxidation, the second describes to biomass respiration, the third describes to nitrification, and the fourth term describes to denitrification.

The demand for oxygen to clean solids from the membrane surface depends on the preceding analysis, and manufacturers suggest a suitable rate of aeration in numerous examples. Three kinds of aerations are utilized in MBR plants: fine bubble aeration, coarse bubble aeration, and rarely jet aeration. (Yu et al., 2006).

1. Design of aeration system

For MBR operation, aeration system is important. In MBR, aeration system in bioreactor bottom must be devised for adequate aeration intensity and immense efficiency transfer to reduce membrane fouling and impede foulant deposition on the membrane surface.Thus, normally, perforation pipe is utilized for aeration approach.To calculate the amount of aeration, three means are: (i) calculate concerning concerning the heuristics gas–water ratio; (ii) calculate according to oxygen transfer efficiency; (iii) or calculate based on bioreactor surface area and aeration intensity.The third method presented for the calculation of aeration amount is carried out.

The determined aeration intensity following the experimental results is 0.01 m3/m2 s.Assuming 5 m of water depth in the bioreactor, the desired amount of air is 247 m3/min and the outer for aeration is 411.4 m2.The selected blowers were four, where three of them were in use and one was supposed for preparation, gas volume recorded with each blower was 82.3 m3/min, power 80 kW, and air pressure 70 kPa.

6.2.2 Operation of Membrane Bioreactor

To control membrane fouling and to optimize system performance is a key role played by operational conditions. Optimized operational conditions and parameters are much favorable in reducing and controlling membrane fouling (Yoon, 2003; Zheng & Liu, 2005; Wang et al., 2006).The crucial concept in submerged MBRs is a critical flux which was focused by various researchers and came up with the results that fouling can be detected above the critical flux, and the fall of flux with time or the gain of transmembrane pressure (TMP) does not appear below the level (Yu et al., 2003; Ognier et al.,2004; Jinsong et al., 2006). Long-term durable operation without repeated cleaning of the membrane could be achieved by MBRs under subcritical operation.The investigation of operational parameters has been done to get better insights of the operational aspects and to optimize these aspects for lessening membrane fouling, and improving the performance; factors include of the ratio of suction, and non-suction time (intermittent filtration), aeration intensity, DO concentration, HRT, SRT, sludge concentration, temperature, and filtration modes (Jin et al., 2006; Zhou et al., 2009).

6.3 PERFORMANCE OF MEMBRANE BIOREACTOR IN WATER TREATMENT

MBR who has treatment efficiency and high effluent water quality is a more competent technology with approving advantage for a commercial company, environment, and society in contrast to conventional water treatment processes.The municipal wastewater treatment is the featured application of MBR, particularly the domestic wastewater treatment. However, for the industrial wastewater treatment, it is also an appealing choice, notably in North America, there are more commercial applications of MBR for industrial wastewater treatment.Moreover, MBR applicable to polluted surface water supply has earned more spotlight. Here, the accomplishment of MBR will be focused on numerous major applications (Shim et al., 2002; Judd, 2008).

6.3.1 Suspended Solids, Organic Matter, and Other Pollutants

Microorganisms decompose most of the organic matters in the MBR systems, and their removal efficiency was embellished by the membrane rejection.Roughly MBR systems, the removal efficiencies of BOD, COD, UV254, and SS are more than 90 percent specifically for the treatment of municipal wastewater.

Nearly all of the suspended solids are taken out in an MBR process. Subsequently, there is also an improvement in the removal of micropollutants adhered to the suspended solids and heavy metals.Formerly, some studies revealed that for the removal of bacteria, the MBR process is remarkably effective.It is exposed by several research works that as compared to classical fecal indicator bacteria, generally viruses are very much resistant to disinfection.Nonetheless, literature has a relatively little depiction and experience referred to as the removal of viruses, which is becoming a bottleneck in recent years.Few studies pinpoint that the virus can be rejected by the membrane, and the removal efficiency can be affected by the size of the membrane pores.Furthermore, the gel layer or the cake layer can also act as a barrier (Beaubien et al., 1996).

6.3.2 Nitrogen Transformation

The better hold of lethargic growing microorganisms (i.e. nitrifies) can be acquired with tremendous biomass concentrations.MBR has strengthened nitrogen removal, and the outcomes are usually decent.

Both anoxic and aerobic stages are required by nitrogen removal processes. Cyclical (off/on) aeration can be a reason for the occurrence of simultaneous nitrification-denitrification (SND) in the frequently fed MBR system.Diffusion drawbacks may produce an anoxic region under inferior DO inside the biological floc where denitrification can show up.Additionally, it is beneficial over traditional nitrogen removal processes if SND is carried out through the shortened way i.e. over nitrites.The preference of SND by the way of nitrite reduced aeration, alkalinity requirements, COD, and cut down biomass yield (Leiknes & Ødegaard, 2007; Le-Clech, 2010).

The determinants that impact SND are chiefly floc size and ambient DO concentration.It is reported that the floc sizes in MBR are smaller than that in CAS regardless of its immense operating MLSS concentration.Until now, the proclaimed SND research works in MBR are totally in anoxic/oxic (A/O) systems functioning under irregular aeration mode.It was noticed that 95% and 83% of the overall nitrogen removal efficiency can be achieved in A/O MBR.Additionally, there has more potential for extractive MBR in the application of nitrogen removal.

6.3.3 Phosphorus Removal

Generally, phosphorus removal is enacted by attending with chemicals, such as lime or metal coagulants, which can constitute lightly soluble precipitates. Anyhow, biological technology is a more economical and environmental technology that does not include supplementary chemicals.Phosphorus is not removed fairly as most of the wastewaters are carbon available treated by biological processes.There might have a slight effect of membrane rejection on phosphorus removal.For biological phosphorus removal, some enhancements have been applied, as an example, the nitrate-free sludge was restored from the aerobic zone and at the front of an activated sludge plant, the anaerobic zone was joined (Feng et al., 2016).

It is shown in Fig. 4 that by using hybrid MBR, the process of phosphorus recovery and denitrification is dependent on MBR technology in which a biofilm carrier being a filling is implanted in the reactor. The anoxic microenvironment is provided in indicated hybrid MBR system which was organized by high concentrated activated sludge Zoogloea and biofilm and suspended. The process of synchronous denitrification and nitrification in the same reactor achieves nitrogen removal. At the same time, to achieve phosphorus release, phosphorus-rich sludge is transferred to the anaerobic zone by extraneous cycling, and afterward crystallization or chemical precipitation reuses the phosphorus (Trussell et al., 2006).

6.3.4 Mechanisms of Virus and Pathogenic Bacteria Removal

For the removal of viruses and bacteria, using the key influence of the cake layer (i.e. irreversible fouling) MBR is inspected. The inactivation of viruses and bacteria in the solid-state is higher than in the liquid state due to a greater density of predators and enzymes. The connection of viruses and bacteria to biomass is too advantageous to removal on the account of operation at a longer residence period and higher MLSS concentrations upgraded the removal of viruses and organisms. The key strategies are given as (1) retention of viruses and bacteria by the cake layer assembled on the membrane surfacewhile eternal operation; (2) inactivation of viruses and bacteria by a chemical improved backwash; (3) viruses and bacteria which are lasted by the membrane are stuck to the MLSSs; and (4) enzymatic or predation breakdown inactivate viruses and bacteria due to the protracted SRT and HRT (Chaudhry et al., 2015).

6.3.5 Drug Residues Elimination

In the MBR system, the efficiency of the antibiotic is less than 60%. In comparison with the effluent, analgesics including tramadol, indomethacin, and diclofenac and nonsteroidal anti-inflamatory drugs incorporated with the influent are relatively unchanged. Whereas paracetamol (acetaminophen), mefenamic acid, metamizole metabolites, and morphine are smoothly biodegradable and after the MBR treatment, more than 92% of the concentrations are minimized. Diuretics, Beta-blockers, and the verapamil metabolite D617 are removed with not well efficiencies i.e. below 55%. There is around 80% of the removal of the other cardiovascular system preparations (verapamil, valsartan, and bezafibrate). The 56% of lidocaine, 91% of anesthetics thiopental, 71% of H2- receptor antagonist ranitidine, and 90% of the enzyme inhibitor cilastatin are eradicated in the MBR (Kovalova et al., 2012).

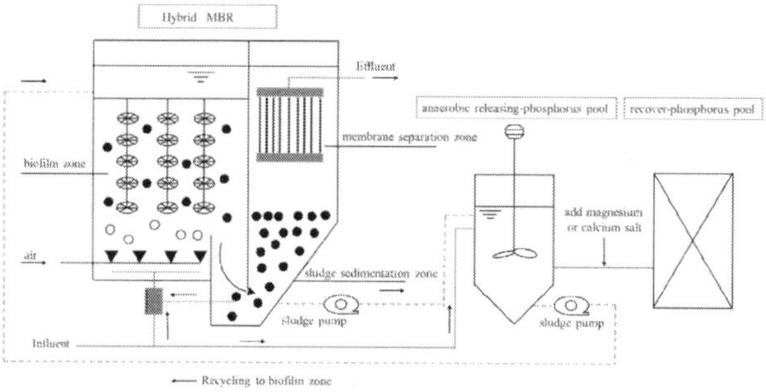

Figure 6.4: For phosphorus and nitrogen recovery procedure, extra-cycle sludge-hybrid MBR.

[Source: https://www.researchgate.net/publication/316141914_Membrane_Bioreactor_Application_in_Water_Treatment]

6.3.6 Special Bacteria Remove Special Contaminants

They are frequently spotted in the effluents of WWTP, groundwater resources, and surface waters because of the world-wide exploitation of drugs again and again. In the recent research work, more than 99.9% of the acetaminophen (APAP)has been eliminated when a 100 mg APAP L1 of synthetic WWTP effluent was provided to the MBR system, where APAP is a type of antipyretic and analgesic drug usually found in water bodies (Howell et al., 2004). No effluent matrix influence in the middle of the processes, no biomass sorption, and the APAP removal with the capacity of the microbial consortium has been proven by the examinations. There were two APAP degrading strains taken from the MBR system i.e. Pseudomonas aeruginosa and Delftia tsuruhatensis. These conclusions reveal that the enhancement of few of the microbial consortiums having particular competency in an MBR would be an auspicious mechanism to deal with the effluents encompassing definite pharmaceuticals (De Gusseme et al., 2011).

6.3.7 MBR for Industrial Wastewater Treatment

At the beginning of 1980s, the process of MBR for the treatment of industrial wastewater is examined by system suppliers of North America and by a small number of researchers.3 MBR processes have been found famous in industrial wastewater treatment from the expanded researches and commercial applications for industrial wastewater treatment, especially in refractory and toxic wastewater, but still, the scale obtains smaller indeed (Wang et al.,

2008). For industrial wastewater treatment, the external configuration is the favorable kind of MBR, which is easily disassembled and cleaned. Numerous industrial wastewater streams enforced several MBR systems which include petrochemical wastewater, food-processing wastewater, dyeing and printing wastewater, hospital wastewater, slaughterhouse wastewater, etc.

Despite some identical characteristics with the municipal wastewater, there are additional special properties of industrial wastewater consists of some exceptional pollutants such as micropollutants, heavy metals, etc or more difficult disposal pollutants. For these pollutants, higher removal efficiencies were shown by the MBR system in some real cases as compared to shown in the CAS system (Moslehiet al., 2008; Llop et al., 2009).

To remove pollutants from petrochemical wastewater, the MBR system was considered to be very efficient. A higher of 90% can be achieved for the removal of both COD and TOC while using MBR for olefin process wastewater treatment. Whereas for complex petrochemical wastewater, the removal of TOC was recorded up to 92%, and the removal of COD was reported up to 83%. For heavy metals, the removal rate achieved of zinc can be up to 60%, chromium can be up to 95%, and lead can up to 62%. For endocrine disrupter chemicals biodegradation, an appropriate environment is provided by MBR for the maintenance of a more varied microbial culture and thorough solids retention in the bioreactor (Clara et al., 2004).

6.3.8 MBR for Drinking Water Treatment

A more integrated technique is offered by MBR, therefore for polluted water supplies, a productive water treatment process can be reasonably developed. The traditional water treatment operations could be consolidated into a single unit by the MBR, counting flocculation, coagulation, sedimentation, disinfection, and filtration. The MBR could be combined with certain processes for treating the water of unique quality, like bio-activated carbon, advanced oxidation processes (AOPs), powdered activated carbon (PAC), etc. Mentioned applications are recently systematically investigated even though this subject has limited literature (Liu et al., 2003).

6.3.9 Organic Carbon and Nitrogen

In consequence of uncritical management and supervision, and economic development, without satisfactory treatment, industrial and domestic wastewaters have been released into natural water-bodies, which results in severe pollution condition of the surface water provides in some regions, with ammonia nitrogen (NH3–N) and organics as the key pollutants (Tian et al.,2009). NNitrate may transmit into drinking water sources since it is water-soluble and does not restrict with soil, and globally it may be possible that nitrate at partial groundwater utilized for drinking water surpassed the utmost contaminant extents. Thus, the pollutant and nitrogen removal turn into stringent for the protection of water supply, and MBR must make use of its benefits on drinking water treatment as a promising and creative process.

Some of the conclusions from the literature of this field sound to be controversial. In submerged MBRfor dealing with simulated polluted surface water, Li & Chu (2003)accomplished 95% of ammonia removal and up to 60% of TOC removal; Tian et al. (2009) attained around 90% ammonia removal and below 50% TOC removal; still, poor performance was reported by some of the researches of MBR applied in this case. For that reason, MBR for drinking water treatment is not considered to be balanced and can be improvised with integrated technology. There is approximately 100% of the performance of MBR for the removal of turbidity which is quite good. For drinking water denitrification, the application of MBR is merely at the development and research phase.2 Similar to MBR wastewater treatment, a new extractive MBR for drinking water treatment has the capability to beat the drawbacks of typical biological denitrification systems.

6.3.10 Micropollutants

AOP can effectively eliminate micropollutants, thus the security of water supply can be ensured by combining MBR with AOP. Williams and Pirbazari (2007)concluded that MBR was impressive when linked with PAC and ozone for the removal of trihalomethane precursors and biodegradable organic substances. Li & Chu (2003)gained almost 75% reduction of the 3-day trihalomethane composition competency with MBR treatment, and the biostability of the effluent upgraded significantly as the assimilable organic carbon (AOC) reduced by 80%. Yet, the outcome of Tian et al. (2009) for removing the AOC was not very satisfactory.

Furthermore, ideal for concerning oxidized compounds in drinking water is the membrane biofilm reactor (MBfR) which adopt hydrogen as an electron donor. Nerenberg and Rittmann (2004) inspected a hydrogen-based, hollow fiber MBfR for the contraction of chlorate, perchlorate, chlorite, chromate, bromate, selenate, arsenate, selenite, and dichloromethane, and the optimal removal efficiency can attain 495%, 498 %, 475%, 475%, 495%, 67%, 450%, 93%, and 38%, respectively.

In the treatment of industrial and municipal wastewater, the efficiency of MBRs has been shown. In wastewater treatment research, the advancement of this process is continuing, and for the technology, providing modeling tools is important.

6.4 CAUSE AND CONTROL OF MEMBRANE FOULING

There are many advantages of MBR in contrast with ordinary wastewater treatment systems. It facilitates with efficient removal of high pollutants. Much higher quality effluent can be produced by MBR while holding soluble microbial products (SMPs) in the system. However, the major barrier to the extensive use of MBR is the membrane fouling. In addition to that MBR users highly concerns about the successive membrane cleaning and related expenses. Membrane fouling has been connected to supernatant characteristics, intermittent operation, and sludge concentration (Huang et al., 2011).

It is important to be familiar with the fouling mechanisms to control the fouling, mechanisms involving the properties of the constituents that induce the membrane fouling in MBRs and the elements influencing the fouling.

6.4.1 Fouling Cause of Membrane in MBR

6.4.1.1 Mechanism of Membrane Fouling

Membrane fouling is an obstacle to broad MBR implementation. The declination of permeate flux at the conditions of fixed trans pressure could refer to as membrane fouling. In MBR, the membrane fouling is attributable to pore blocking, membrane pore constriction, and cake layer formation by colloids, solutes, and sludge particles as shown in Fig. 6.5 (Rosenberger et al., 2006).

The increase in the fouling rate is nearly exponentially operating above critical flux, so usually, MBR is regulated under critical flux conditions. Although indeed subcritical flux operation can accelerate fouling corresponding to a two-stage pattern: a low TMP grows above an initial period pursued by a rapid rise after some threshold period, except for that an initial limited rapid increase in TMP is noticed. Therefore, a three-stage fouling process is presented(Chang et al., 2002; Zhang et al., 2006):

Stage 1: an initial momentary rapid growth in TMP;

Stage 2: a longstanding weak increment in TMP;

Stage 3: an abrupt increase in dTMP/dt, which is also known to be TMP jump.

The representational diagram of the TMP jump is demonstrated in Fig. 6.6. It is considered to be the aftereffect of severe membrane fouling. The TMP jump is associated with the fluctuations in the local flux as a result of fouling by Cho and Fane (2002), which is the reason for higher local fluxes than the critical flux. It is also reported by Zhang et al. (2006)that local flux was not the only reason for the immediate jump, aside from it can also be provoked by sudden transitions of the cake layer structure or biofilm. Due to insufficient transfer of oxygen to the biofilm, biofilm had released more EPS (Cho & Fane, 2002; Hwang et al., 2008).

The membrane fouling appears due to the following mentioned mechanisms relative to MBR:

1. Arrangement of a cake layer on the membrane surface;
2. Adsorption of colloids or solutes inside the membranes;
3. Disconnection of foulants attributed chiefly to shear forces;
4. Removal of sludge flocs upon the membrane surface;
5. During the long-term operation, the temporal and spatial revisions of the foulant composition.

6.4.1.2 Species of Membrane Foulants

The membrane fouling can be described as the unwanted dismissal and aggregation of colloids, cell debris, and microorganisms on the surface of the membrane or within membranes.

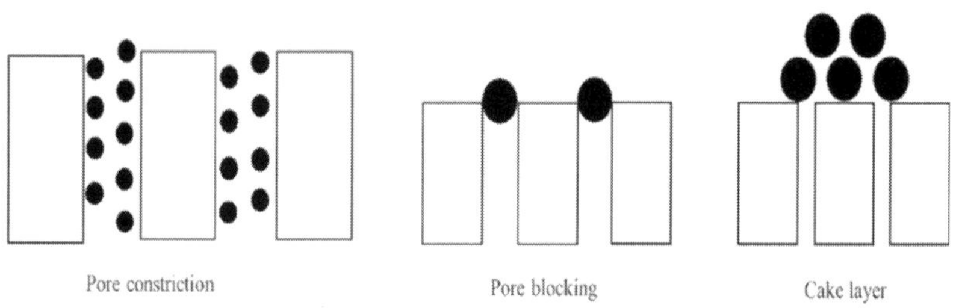

Figure 6.5: MBR'smembrane fouling.

[Source:https://www.researchgate.net/publication/316141914_Membrane_Bioreactor_Application_in_Water_Treatment]

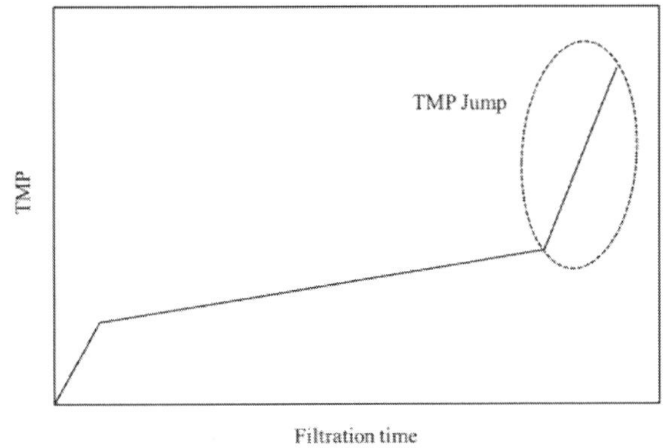

Figure 6.6: Representational diagram of the TMP jump.

[Source: https://www.semanticscholar.org/paper/PERFORMANCE-OF-ANAEROBIC-MEMBRANE-BIOREACTORS-WITH-Wong-Chong/ee-68a1007a2514a858201ed3914ae021c165147f/figure/1]

As the complicated nature of the activated sludge is known, so it is not the astonishing fact that the fouling act in MBR is more intricated as compared to that in many membrane applications. In MBR, there are two categorizations of membrane fouling: (i) physically irreversible fouling which physical membrane cleaning cannot possibly remove, and (ii) physically reversible fouling which physical membrane cleaning (such as surface cleaning or backwashing) can eliminate. Even though the focus of nearly all of the recent studies have on the means of membrane fouling, but the extremely valuable work for maintaining sustainable and long-term operation and minimizing the operation cost of MBR is the control of irreversible fouling. The fouling can be categorized into three leading classes concerning the fouling factors: organic fouling, inorganic fouling, and biofouling (Liu et al., 2012).

The deposition of biopolymers (polysaccharides and proteins) on the membranes is called organic fouling (primary focus on SMPand EPS) in MBR. A detailed study is conducted by Metzger et al. (2007) to distinguish accumulated biopolymers in MBR. The fouling layers were portioned into the upper layer, lower layer, and intermediate layer after membrane filtration by using rinsing, chemical cleaning, and backwashing. The outcomes illustrated that the upper fouling layer was constituted of a porous, freely bound cake layer with the same formation to the sludge flocs. The lower layer which is noticeable by SMP and depicting the irremovable fouling ratio had a comparative immense concentration of bound proteins. The intermediate fouling layer was assisting uniformly by bacteria aggregates and SMP and had a higher accumulation of polysaccharides. On the membrane surface, the spatial dissemination of biopolymers is exposed by this study. The important examining tools for the inspection of organic fouling are: ultraviolet-visible spectrometry, Fourier transform infrared spectroscopy, excitation-emission spectroscopy, high-performance size exclusion chromatography, and solid-state 13C-nuclear magnetic resonance spectroscopy. It is approved by these studies that EPS or SMP is the inception of organic fouling, and it portrays an essential part in the progress of MBR fouling (Song et al., 2003).

A little much paper has referred Inorganic fouling. The filtration properties of inorganic and organic membranes were explored by Kang et al. (2002)in a membrane-coupled anaerobic bioreactor, to which a dense cake layer composed of the membranes (particularly on the inorganic membrane surface) which is made up of struvite and biomass. The formation of a cake layer is strengthened by the organic foulants joined with the inorganic precipitation. These discoveries signify the increasing importance of inorganic fouling in MBR. Except for that, the perception of inorganic fouling is even not sure. Therefore, it is an up-and-coming research problem (Kornboonraksa & Lee, 2009).

Biofouling specifies the growth, deposition, and flocs on the membranes or metabolism of bacteria cells, which has inflamed a symbolic interest in membrane filtration processes (Pang et al., 2005). It is a significant problem for wastewater treatment for a low-pressure membrane e.g. ultrafiltration and microfiltration, due to the much larger most of the foulants in MBR as compared to the membrane pore size. Biofouling is deduced from two circumstances, one is the cells enlarging and composing a biocake, and the other is the cell group on the membrane surface or the deposition of an exclusive cell. Several analysts advised that EPS and SMP secreted by bacteria has outstanding roles in the cake layer on membrane surfaces and the establishment of biological foulants (Flemming et al., 1997). They also proclaimed that the microbial associations on membrane surfaces were distinct from the ones in the suspended biomass. In the production of mature biofilm, the beta-proteobacteria did a major job, which concluded to severe irremovable membrane fouling. However, it is reported by Jin et al. (2006) that the most ambiguous species is the g-proteobacteria as compared to other microorganisms, whereas the higher surface hydrophobicity belongs to the deposited cells in contrast to the suspended sludge (Tay et al., 2007).

6.4.1.3 Seasonal Variation in Membrane Fouling

Recent studies have been worked on the dissimilarity of membrane fouling and seasonal variation. During the operational processes of MBR, seasonal variations of membrane fouling were recognized expecting physically irreversible and reversible fouling. It is observed that in the MBR with short SRT, physically irreversible fouling matured with more speed in the period of high-temperature, whereas physically reversible fouling came into view more clearly in the period of low-temperature. The logic behind seasonal variation in physically reversible fouling is associated to change in the quantity of organic matter, and irreversible fouling is correlated to variations in the quality of organic matter (Miyoshi et al., 2009).

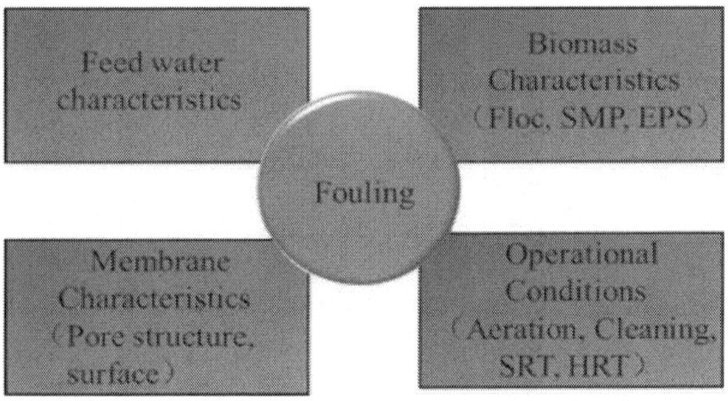

Figure 6.7: Aspects of leading membrane fouling.

[Source:https://www.researchgate.net/publication/316141914_Membrane_Bioreactor_Application_in_Water_Treatment]

6.4.2 Fouling Control of MBR

There is an impact of all the attributes participating in the operation and design of MBR processes on membrane fouling. But the degree of membrane fouling is firmly affected by four parameters (Fig. 6.7): (i) biomass characteristics, (ii) membrane characteristics, (iii) operational conditions, and (iv) feedwater characteristics (Le-Clech et al., 2006).

During the long-lasting operation of MBRs, TMP jump appears unavoidably under the fashion of constant flux. Accordingly, to lessen the appearance of the TMP jump is the general target of fouling control.

Numerous means had been recommended to alleviate/control membrane fouling depends on the parameters which prompt the membrane fouling are as follows (Ersu et al., 2010; Di Bella et al., 2015):

1. Pretreatment of feedwater

The biological, chemical, and physical properties of feedwater can be altered by pretreatments of feedwater which makes better the MBR's process performance. Huang et al. (2009) had summarized three mechanisms that are as follows:

- i Biological mechanisms: Pretreatment can eliminate biodegradable compounds appropriate to decontaminate feedwater to cut down biofilm formation, or treated-water quality or membrane fouling.
- ii Chemical mechanisms: To shorten the affinity of foulants to membrane surfaces and alter water chemistry, different chemicals (oxidants, coagulants, and adsorbents) can be combined with the feedwater, by that relieving irreversible membrane fouling.
- iii Physical mechanisms: The size of aquatic material can be increased by pretreatment to the extent breakable by the membrane, thereby improving their eliminations by the integrated membrane system. An increase in size also obstructs membrane fouling to cake filtration that is normally more reversible and littler severe or shifting it from pore constriction.

Adsorption, oxidation, and coagulation are the vital pretreatments of feedwater worked at full-scale filtration convenience. The exact description of the mechanism of various pretreatment methods is discussed by Huang et al. (2009)along with their pros and cons.

2. Modification of biomass characteristics

From biomass, secretion, SMP, and bound EPS prevail in the MBR systems and are watched as the preeminent foulants. EPS or SMP can pass through the membrane pores or add on the membranes. The back-transport rate of the hoarded particles from the membrane surface toward the bulk and particle convection approaching the membrane surface demonstrates the detachment and accumulation of membrane foulants. The back-transport method in membrane filtration consists of shear-induced diffusion, Brownian diffusion, and inertial lift.

Furthermore, the modification to biomass has attempted by using ozone, ultrasound, magnetic enzyme carrier, and electric field (Yeon et al., 2009a; 2009b). Membrane fouling could be controlled by ultrasound effectively as shown by experimental results, even though supporting some operation conditions, membrane damage may happen (Wen et al., 2008). Comparing to traditional MBR having no enzyme, the laboratory-scale MBR applied magnetics enzyme carrier in a non-stop operation, and it also raised the membrane permeability to a large degree. Other than that, the use of an electric field is an appealing method that could impede the sludge flocs and colloids invested onto the membrane surface (Chen et al., 2007).

Anyhow, henceforth, further methods might be introduced based on the attributes of biomass.

3. Bacterial quorum quenching for anti-biofouling technique

A rising antifouling method that has recently researched is Quorum quenching (QQ). The growth of TMP relevant to the time was highly alleviated in QQ-MBR in the study than that in the standard MBR. Hence, QQ-MBR can retain the quality of the effluent water and quietly preserve energy consumption. It is due to the QQ-beads are combined into a typical MBR to influence the microbial floc size in the system because of biodegradation capability to signaling molecules as well as, EPS concentrations (Lee et al., 2016).

4. Optimization of operational conditions

A potent approach to stay away from severe fouling is the operation under the critical flux which is inclusive of irreversible and reversible fouling. Non-fouling operation or subcritical flux is the concept of operation under the critical flux. It is expected to bring on small irreversible fouling. The value of critical flux depends on operating conditions (i.e., temperature, aeration intensity), membrane features, and sludge attributes. The study of MBR fouling has widely used the thought of optimization of operating conditions in MBR that may be able to control the critical flux value.

After membrane fouling particularly irreversible fouling has a significant part in the eternal operation of MBRs, every so often to manage operations of MBR chemical cleaning is mandatory. Passing over, chemical cleaning for the reduction of irreversible fouling must be restricted to the least frequency because of the reason that the disposal of wasted chemical agents would lead to environmental issues and recurring chemical cleaning might short down the lifetime of membrane.

There are three leading roles of aeration adopted in MBR systems that are: to diminish fouling by the consistent scouring of the membrane surface, to keep up the activated sludge in suspension, and to supply oxygen to the biomass. A thorough investigation and review have been done especially for MBRs on the practice of gas bubbling to upgrade membrane processes (Hong et al., 2012).

Moreover, there are variant influences of all of the HRT, F/M (food-to-microorganism), SRT, and alternate operational attributes on membrane fouling of MBR.

5. Exploration of new membrane materials

The impact of membrane characteristics like porosity, pore size, roughness, surface charge, hydrophobicity\hydrophilicity, etc. have been proven on MBR achievement, exclusively on membrane fouling. Diverse fouling property is shown by the membrane materials perpetually due to their variant morphology, pore size, and hydrophobicity. To prevent irreversible fouling of MBRs used for the operation of municipal wastewater, PVDF membrane is preferable to PE membrane. The order of the affinity capacity of the three membranes was: PANoPVDFoPES. It recommends that out of these mentioned membranes, more fouling resistance is the PAN membrane (Ngo & Guo, 2009; Xiao et al., 2019).

As the hydrophobic corporate between membranes and foulants, so generally membrane fouling appears more quickly on hydrophobic membranes as compared to on hydrophilic ones. Consequently, considerable attention has been delivered to decrease membrane fouling by transforming hydrophobic membranes into corresponding hydrophilic membranes (Tsai et al., 2005; Holloway et al., 2015).

6.5 MODELS OF MEMBRANE BIOREACTOR

Notably in water treatment systems, as a result of the popularity and emergency of MBR, for numerous membranologists, simulation and correct prediction of the entire MBR process are becoming noteworthy and ubiquitous, were submerged and external configuration is incorporated by MBR for membrane modules, (Cho & Fane, 2002; Wang et al., 2014; 2016). The factual measurement and effective design of MBR could bring MBR to light. Normally, four categories are being classified by these models: (1) organics removal models, (2) biomass kinetic models,(3) membrane fouling models, and (4) integrated models along with the above pairings to express the complete MBR (Ng & Kim, 2007).

6.5.1 Organics Removal Model

The kinetic tendency of organic substrate declination in an SMBR system is disparate from that in the CAS system due to the ensuing causes. Firstly, nearly all microbes can be kept in the bioreactor by MBR, which results in the holding of high sludge concentration, hence there is an extremely low organic-sludge load. Plus, due to low organic-sludge load and high sludge concentration, the organic substrate is degraded by the microorganisms of the reactor principally to control their activity. Furthermore, in the reactor SMP can also be retained by MBR, which is vitiated additionally as an organic substrate. Respective of the practical attributes of submerged MBR, to provide a reference for the scheme of the SMBR system, the model must be necessarily established for discharging organic substrate (Chen et al., 2006; Drews, 2010).

In submerged MBR, Fig. 6.8 shows the balance of material. For the organic substrate in the reactor, Eq. (3) is practiced to elaborate on the mass balance in accordance with Fig. 6.8 #

$$Q_0 S = Q_e S_e + Q_r S_r + U_y V + V\frac{dS_r}{dt} \qquad (3)$$

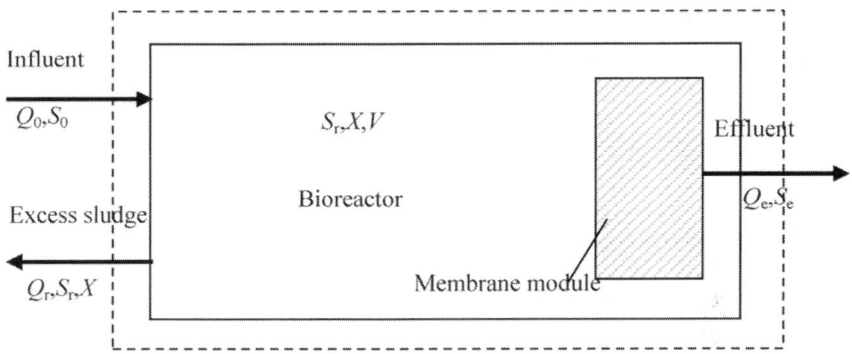

Figure 6.8: Balance of material in SMBR.

[Source:https://www.researchgate.net/publication/316141914_Membrane_Bioreactor_Application_in_Water_Treatment]

Where

Q0 depicts the influent flow rate (m3/day),

S0 indicates the influent organic substrate concentration (mg/L),

Qe signifies the effluent flow rate (m3/day),

Se point out the effluent organic substrate concentration (mg/L),

Qr marks the surplus sludge discharging rate (m3/day),

Uy represents the removal rate of organic substrate (mg/L day),

V symbolizes the productive volume of the bioreactor (m3), and

dSr/dt shows the organic substrate variation rate in the reactor (mg/L day).

On a steady-state condition of the system, the reactor's organic substrate variation rate can be ignored. Related to Eq. (4) and Eq. (5), Eq. (3) is equal to Eq. (6), and Eq. (6) can be altered to Eq. (7). All these equations are presented as follows:

$$Q_0 = Q_e + Q_r \quad (4)$$

$$S_r = S_e + S_m \quad (5)$$

$$Q_0(S_0 - S_e) = Q_r S_m + U_y V \quad (6)$$

$$U_y = \frac{Q_0(S_0 - S_e)}{V} - \frac{Q_r S_m}{V} \quad (7)$$

where Sm signifies the organic substrate concentration cleared away by membrane module (mg/L).

From the subsequent replacement

$$\text{HRT} = \frac{V}{Q_0}, \text{SRT} = \frac{V}{Q_r} \quad (8)$$

where HRT interprets the wastewater's hydraulic retention time in the reactor whereas SRT represents the sludge retention time, Eq. (7) can be converted to Eq. (9), that is to say, the model of cutting out an organic matter in consideration of the SMBR system, and abbreviated as ROM

$$U_y = \frac{S_0 - S_e}{\text{HRT}} - \frac{S_m}{\text{SRT}} \quad (9)$$

Concerning the model introduced by Wisniewski et al., for an SMBR system, only the HRT of the reactor and organic substrate concentrations of the effluent and influent determines the organic substrate removal rate in steady state. The concept of Wisniewski's model is that the organic substrate concentration remaining in the reactor and that of the effluent is same, aside from that several researches have been done that concluded that the organic substrate concentration of the effluent and that endured in the reactor is different, to the extent that the disparity is substantial (Meng et al., 2005; Lin et al., 2014). Thus, the conclusion can be drawn from the ROM model that not only the HRT of the reactor and the organic substrate concentrations of the effluent and influent concluded the organic substrate removal rate, except for that also the SRT participated at the steady-state of the SMBR system and the organic substrate concentration ejected by membrane modules (Meng et al., 2009).

6.5.2 Biomass Kinetics Models

Biomass kinetics models comprise of SMP model, activated sludge model (ASM) family, and the hybrid of SMP and ASM models (Wintgens et al., 2003).

The prominent dissimilarity between MBR and CAS biological system is the partition module of the activated sludge and effluent, which acts as the membrane for the MBR separations at the time of secondary clarifier for traditional biological treatment separation. Thus, in the MBR system, ASM family designing CAS systems is convenient for the biomass kinetics (Verrecht et al., 2008; Naessens et al., 2012).

According to the International Water Association (IWA), the ASMs amount to (i) ASM no.1, (ii) ASM no.2, (iii) ASM no.2d, and (iv) ASM no.3. Firstly, ASM no.1 was proposed in 1987 for the design and working of the biological wastewater treatment process. The others introduced in the later years were the improved and expanded version of the first model: ASM no.2 organized to the removal of the phosphorus from wastewater; ASM no.2d evaluate the competency of phosphorus-accumulating organisms to utilize cell internal substrates concerning denitrification whereas ASM no.3 addressed all the problems found except for phosphorus removal.

Hence, the four ASM models have their separate focal points. The simulation of the process of organic matter removals, denitrification, and nitrification in the activated sludge system is commonly associated with ASM no.1 in which heterotrophs acts as denitrifiers and autotrophs serve as nitrifiers. ASM no.2 is competent to accumulate phosphorous into ASM no.1 in the pattern of internal cell materials and assimilate a group of organisms. The phosphorous-accumulating organisms (PAOs) are capable to mold the external soluble phosphates into polyhydroxyalkanoates (XPHA) or polyphosphates (XPP) of internal cells materials. At aerobic conditions, the raise of PAOs is based on the consumption and storage of XPHA, although the energy for XPHA storage is afforded by the hydrolysis of XPP. In comparison, under anaerobic conditions, the growth of PAOs is illustrated by ASM no.2d, at the same time the denitrification is encompassed. In the ASM No.3 model, the storage is focused on heterotrophs instead of hydrolysis of organic substrates, therefore external compounds do not determine the

CHAPTER

7

AEROBIC MEMBRANE BIOREACTORS FOR WATER TREATMENT

CONTENTS

7.1 INTRODUCTION

In wastewater treatment, aerobic membrane bioreactors called MBRs are a prime technology in achieving sustainability through decentralization, low energy consumption, and reuse (Fane and Fane 2005; Fawehinmi et al. 2005). In these MBRs, utilizing membrane processes, the aerated activated sludge is combined with membrane process for separating the solids from industrial or treated municipal wastewater and for removing dissolved contaminants of ammonia and carbon. Microorganisms remove carbon and metabolize it in access of dissolved oxygen for respiration and microbial growth. Ammonia oxidation removes ammonia (nitrification). Nitrification is a process in which microbial mitigated reduction takes place in an aquatic medium containing medium to high ammonia concentration along with comparatively lower organic carbon concentration and oxygen in dissolved form.

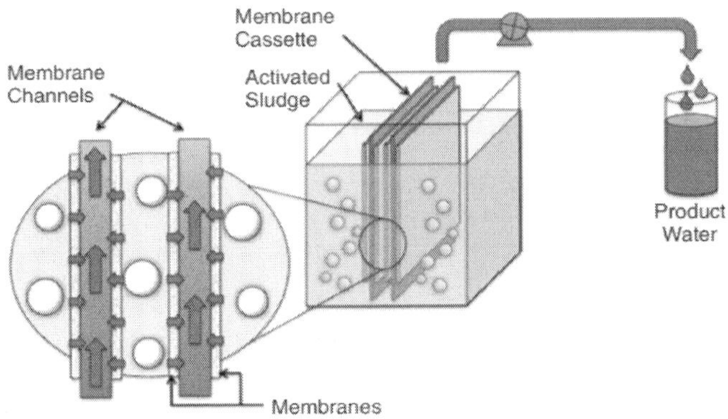

Figure 7.1: Schematic presentation of an aerobic MBR system(submerged).

[Source:-https://link.springer.com/referenceworkentry/10.1007%2F978-3-642-40872-4_7-1]

Fig. 7.1 presents submerged MBRs. In a bioreactor, the microporous membranes (ultrafiltration (UF) or microfiltration (MF) (▸Ultrafiltration)) are immersed and through the membranes, water is filtered (▸Permeate) using vacuum; in the system, solids suspended are retained and through it the high treatment level might be attained (Judd 2006). The conventional process of two-stage activated sludge (clarification and biotreatment) is replaced by the integrated single process i.e. MBR (▸Wastewater Treatment in Membrane Bioreactors). Thoroughly, the benefits MBR has over the conventional treatment are reviewed (Stephenson et al. 2000); reduced footprint, product consistency, separation of solids from the effluent, and reduced sludge production has been included. Additionally, for utilizing water from irrigation as processed water or utilizing it for potable pretreatment to reuse applications, MBR may be more affluent (►Membrane Bioreactors for Reuse; ►Potable Water Production) (Lawrence et al. 2002).

Nonetheless, establishing technology of membrane bioreactor (►Growth Membrane Bioreactor Attached; ►Submerged Membrane Bioreactor; ►Membrane Bioreactors (MBR) used for Plants; ►Anaerobic Membrane Bioreactor) are slower than presumed. This is due to the reason that decision-makers consider MBRs highly expensive and great risk in contrast to conventional technology (Judd 2006). Till today, MBRs are utilized in treating industrial and municipal wastewater for reusing it, a minor footprint is needed, or stringent standards of discharging exist (Kang et al. 2007; Yang et al. 2006).

7.2 CURRENT LIMITATIONS

Controlling the fouling of membrane through chemical input and modest energy (Reversible Fouling; ►Irreversible Fouling) is a major constraint faced in wide-spreading the technology of MBR (Le-Clech et al. 2006). The fouling of membrane (►Cake Layer; ►Biological Fouling; ►Inorganic Scaling) is investigated from different perspectives which include the mechanism, characteristics, and causes of fouling, to reduce them or to prevent them, methods have also been discussed (Achilli et al. 2011; Judd 2005; Le-Clech et al. 2006; Meng et al. 2009; Wang and Wu 2009). Fouling noticeably affects the productivity of the system, cleaning of a membrane (Membrane cleaning, 369690), the integrity of membrane and replacement interval; these all factors affect the costs and requirements of energy (Judd 2006; Le-Clech et al. 2006).

7.3 OPERATION

At constant flux, for operating conventional MBRs, the cleaning techniques for physical membranes are used; these techniques include relaxation, backwashing, air scouring, or all three combined, depending on the configuration of a membrane (flat sheet, hollow fibers or tubular). For MBR having submerged configuration, air scouring is needed so that fresh sludge can be gas lifted via membrane cassette or bundle and so that solids can be scoured through the surface of the membrane. In the backwashing technique, via the membrane, permeate is pumped towards the opposite direction, this efficiently removes the majority of the reversible fouling A detailed study of the effectiveness of this backwashing technique is done and the parameters that are concluded essentially are duration, intensity, and frequency (Bouhabila et al. 2001; Psoch and Schiewer 2005, 2006). The permeate suction is ceased during the technique of membrane relaxation and naturally, foulant's back transport is amplified; this is because the foulants that are reversibly attached, start diffusing away from the surface of the membrane.

To control fouling in hollow fiber and tubular membranes, the regularly used techniques are of membrane relaxation and backwashing (Bouhabila et al. 2001; Hong et al. 2002; Psoch and Schiewer 2005, 2006; Smith et al. 2005). Some membranes may not be backwashed, they are called flat-sheet membranes as they are not able to hold out the pressure applied in the conventional direction of flow; for controlling the membranes fouling, relaxation is used (Le-Clech et al. 2006). At regular intervals, for enhancing physical cleaning, chemicals must be utilized in any case of membrane configuration (Le-Clech et al. 2005).

7.4 CONTEMPORARY STATUS OF AEROBIC MBRS

7.4.1 Aerobic MBRs used in Markets for Commercial Purposes

Over 200 countries have installed MBR systems, in accordance with Grand View Research (GVR) report, in 2016; globally, the size of the MBR market was valued at 1.81 billion USD. As global interest in wastewater treatment is increasing, by 2025, the market may reach to 8.27 billion USD. According to a report by BBC, from 2018-2023, 14.7% will be the compound annual growth rate (CAGR) and it will make the global market jump from 1.9 billion USD in 2018 to 3.8 billion USD in 2023 (Mateo-Sagasta t al., 2015; Mert et al., 2018; Zhang et al.,2020).

Asia-Pacific was presiding in 2016 and 50.4% of the global market of membrane bioreactor was accounted for by it. Another significant market was North America and 28.0 % overall industrial share was accounted for by it. From the value of 769.2 million USD in 2018 to 1.8 billion USD by 2023, the MBR market for Asia-Pacific should grow and in this period (2018–2023), the CAGR would be 18.5 %. Similarly, from the value of 510.3 million USD in 2018 to 957.0 million USD by 2023, the MBR market for North America should grow and in this period (2018–2023), the CAGR would be 13.4% (Shannon et al., 2010; Ang & Mohammad, 2019).

In 2016, accounting 80.7% of the international market, hollow fiber was presiding configuration. It is expected that this configuration will show efficient growth as operational and maintenance costs are comparatively lower than a flat sheet and tubular configurations. It is expected that hollow-fiber membranes demand may increase in the forecast period because of high adoption within municipal installations that are large-scaled (Daigger, 2014; Xiao et al., 2019).

In municipal wastewater, membranes use is increasing as high packaging density and productivity per unit volume is provided. Additionally, it is expected that the demand for membrane may increase with the growing populations of China, Bangladesh, India, Mexico, Thailand, and Turkey. Within the Asia Pacific, China is the mega MBR growth market; numerous MBRs (especially large-scaled ones) are installed in plants used for wastewater treatment (Judd, 2010; Zgavarogea et al., 2016).

Five-Year plans by the government have driven this activity; by making water reuse mandatory. Many local companies (membrane) like Memstar and Beijing Origin Water (BOW), completely dominate due to the reason that they are aligned with those companies that build wastewater infrastructure. If awareness about proper sanitation is increased, as an outcome, more wastewater and urban water treatment facilities will be developed in other countries like India, Mexico, and Brazil.

However, MBR adoption will be limited by high operating costs and great initial investment. An enormous industrial potential will be created by technical developments, done to reduce footprint, reduce sludge production, and to increase output (Wanner, 1994; Jenkins et al., 2003).

The global industry of membrane bioreactor is very competitive and dynamic; this is due to the presence of numerous large and small manufacturers. The predominant players are Kubota Corporation, Evoqua Water Technologies LLC, SUEZ Water Technologies & Solutions, Koch Membrane Systems, Inc., Mitsubishi Rayon Co., Ltd, and Toray Industries, Inc.

In the coming time, MBRs will surely gain momentum as it is an environment-friendly and economically viable technology for treating wastewater. It provides benefits like a small footprint, outstanding water quality, and high efficiency of biodegradation; these all will expand the MBR industry in upcoming years (Tajima & Yamamoto, 1988).

7.4.2 Strategies for Controlling Membrane-Fouling

The fouling of membrane is affected by the factors that can be classified concerning about the hydrodynamic environment, wastewater composition, sludge characteristics, and membrane properties. Elements like soluble microbial products (SMP), extracellular polymeric substances (EPS), biofloc/biomass, or colloids inside the wastewater undergo interaction with membranes in several ways, design and operation characteristics like dissolved oxygen (DO) concentration, organic load, or HRT and SRT. The properties of the membrane are a fundamental factor in the determination of membrane fouling.

Table 7.1: Large-scaled MBR plants, all around the world (Hollowayet al., 2015; Wang et al., 2016; 2017)

Installation	Location, country	Average daily flow (m^3/d)	Commis-sioning date	Tech-nology supplier
Huaifang Water Recycling Project	Beijing, China	600,000[a]	2016	CITIC
Henriksdal WWTP	Stockholm, Sweden	536,000[b]	2019	SUEZ
Luo Fang WWTP	Shenzhen, Guangdong, China	400,000	2018	SUEZ
Beihu WWTP	Hubei, China	400,000[c]	2019	BOW
Guizhou Xingyi Phase 1	Xingyi, Guizhou, China	255,675	2016	BOW
Seine Aval WWRP	Ache`res, France	218,000	2017	SUEZ
Macau WWTP	China	210,000	2018	SUEZ
Zhuhai Qianshan WWTP	Guangdong, China	210,000	2016	BOW
Sanjintan WWTP	Wuhan, Hubei, China	203,000	2015	BOW
Huhehaote Xinxinban WWTP	Mongolia, China	203,000	2016	BOW
Jilin WWTP (Phase 1, upgrade)	Jilin Province, China	203,000	2015	BOW
Weibei Industrial Park Wanzi WWTP	Xi'an, China	203,000	2016	BOW

Caotan WWTP PPP project	Xi'an, Shanxi, China	203,000	2015–2017	BOW
Liaoyang City Centre WWTP Phase 2	Liaoyang, China	200,000	2012	CITIC
Gaoyang Textile Industrial Park WWTP Phase 1 & 2 & 3	Gaoyang, Hebei, China	200,000[d]	2016	CITIC
Fuzhou Yangli WWTP (Phase IV)	Fuzhou, China	200,000	2015	CITIC
Chengdu Xingrong Project #3	Chengdu, China	200,000[d]	2016	CITIC
Chengdu Xingrong Project #5	Chengdu, China	200,000[d]	2016	CITIC
Chengdu Xingrong Project #8	Chengdu, China	200,000[d]	2016	CITIC
Shunyi WWTP	Beijing, China	180,000	2016	SUEZ

a Only operating at 400,000m3/d. b Commissioning to be done in four phases (2018–2022). c ADF: 400,000m3/d MBR+400,000m3/d conventional activated sludge (CAS). d Only operating at 70% of capacity.

Pore size, hydrophilicity, and membrane chemistry affect initial biomass attachment and permeability; fabrication and configuration of cassettes, modules, trains, and MF/UF membranes are the other factors that affect it too. Additionally, proper hydrodynamic conditions like module design, backwash, and imposed flux can mitigate fouling (Cote et al., 2012).

7.4.3 Economic Analysis and Life Cycle Assessment of Aerobic MBRs

For evaluating the impact of the complete process on environment, service, or product, life cycle assessment (LCA) is utilized; at the policy level, it helps in gaining information. However, LCAs are uncommon for MBR, and pilot studies' finite scale (utilizing only 1m3 of municipal wastewater) affects the performance of energy negatively, this raises questions on the reliability of energy demand calculations. Under varying operating conditions, LCA for MBR should be piloted so that the process can be optimized and environmental assessment can be conducted properly (Zhang et al., 2006; Krzeminski et al., 2017).

7.5 AEROBIC MBRS' RECENT DEVELOPMENTS

7.5.1 Evaluation in Configuration and Membrane for Aerobic MBRs

The betterment in membranes and MBR configurations (aerobic) is due to the driving forces exerted by membrane fouling control, better effluent quality, higher membrane lifespan, and lower operating cost. Membrane development earmarks improvement in the antifouling capacity of membrane or decrement in the cost of a membrane module. Filtration performance and membrane properties are purposive by diverse characteristics like surface charge, roughness, hydrophobicity/hydrophilicity, and pore size (Verrecht et al., 2010; Jeon et al., 2016). Usually, membranes fabricated by different materials hold, unlike properties because of differences in hydrophobicity, foulants interactions, and morphology. For example, polymeric ultrafiltration membranes of polyethersulfone (PES), polyacrylonitrile (PAN) and polyvinylidene fluoride (PVDF) have dissimilar characteristics due to their versatile affinities for EPS and in chemical cleaning, as they differ in resistance (Abass et al.,2015; Ioannou-Ttofa et al.,2016).

Several attempts have been made for developing configurations of the novel MBR system (aerobic). Till today, numerous kinds of novel MBRs systems have been developed and are still being proposed. Fig. 7.2 displays configurational schematics of discussed MBRs. Depending on the purpose, in four groups; these configurations can be classified (1) energy reduction, (2) wastewater reuse, (3) specific foulants removal, and (4) fouling mitigation and enhanced pollutants removal (Luo et al., 2017; Morrow et al., 2018). Most of these MBR configurations are fabricated by the integration of these MBR process along with some other technology or by optimization of their configuration design. For instance, removal of pollutants and mitigation of fouling of HFMBR (Fig. 7.2A), SRMBR (Fig. 7.2D) and AGMBR (Fig. 7.2B) were efficiently enhanced by integrating rotating membranes, aerobic granular sludge, and biofilm technology into MBR process in a particular pattern. By optimization of the design of mass transfer and aeration modes, the potential of fouling in JLMBR (Fig. 7.2F) and AsMBR (Fig. 3E) was highly reduced. Additionally, the union of membrane distillation (MD) and osmotic membranes (OM) along with the process of MBR (MDBR (Fig. 7.2G), OMBR-MD, OMBR, and MBR+RO) helps in achieving effluent of high quality for potable reuse (Tangsubkul et al.,2006; Phattaranawik et al., 2008).

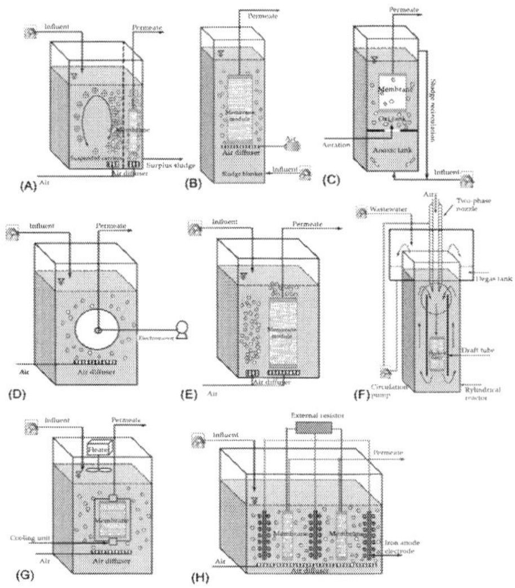

Figure 7.2: Schematic diagram of (A) HFMBR, (B) AGMBR, (C) VSMBR, (D) SRMBR, (E) AsMBR, (F) JLMBR, (G) MDBR, and (H) BEMR.

[Source:http://www.sciencedirect.com/science/article/pii/B9780128198094000012]

7.5.2 Removal of Emerging Contaminants by Aerobic MBRs

Emerging contaminants (ECs) are referred as chemicals that exhibit the potential of posing risks to the environment or human health, these chemicals are still not been subjected to the protection norms or regulatory criteria of environment or human health (Wang et al.,2011; Sauvé & Desrosiers, 2014). These chemicals include pollutants like pharmaceuticals, endocrine-disrupting compounds (EDCs), personal care products (PPCPs), fuel additives, pesticides, plasticizers, and many more industrial compounds (Meng et al., 2012; Jhaveri & Murthy, 2016). Among all ECs, the highest visibility is received by the pharmaceutical substances as they are tenacious in the environment, they may promote the proliferation or development of bacterial strains and as their effects are chronically toxic (Ong & Hu, 2008; Mailler et al., 2013; Li et al., 2015). Recent studies on ECs had more focus on the investigation of pharmaceutical substances. Removing ECs from conventional wastewater treatment plants (WWTPs) cannot be achieved as they are not designed for this purpose, but the emission of ECs can be reduced. Hence, in aquatic environments, WWTPs are one of the greatest

ECs emission sources. As ECs will adversely affect the environment and human health in the aquatic environment, it is important to find out appropriate techniques for removing ECs (Gómez-Espinosa & Arizmendi-Cotero, 2017).

7.5.3 Reduced Energy Consumption in Aerobic MBRs

Compared to conventional activated sludge (CAS), MBRs utilizes greater energy as its aeration requirements are more. They need aeration for both, membrane and biology (Fig 7.3). Biological aeration is needed as it is responsible for supplying oxygen to microorganisms for organic degradation. The membrane aeration is needed as it removes the solid and prevents the membrane from fouling from its surface. For maintaining optimal membrane performance (low fouling and high flux), it is necessary to supply enough air to the surface of the membrane. In the 1990s, it was found that without having any effect on membrane performance, energy can be reduced through intermittent membrane aeration (Dolar et al., 2012).

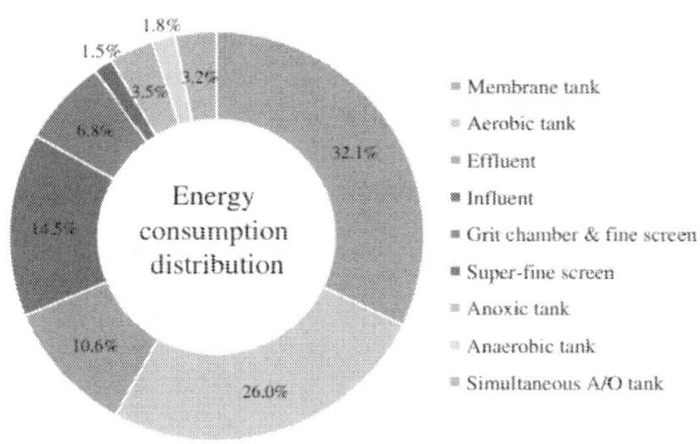

Figure 7.3: Energy consumption constituting full-scale MBR since 2012.

[Source:http://www.sciencedirect.com/science/article/pii/B978012819 8094000012]

Later on, in the 2000s, with the technique of undefined cycle, the intermittent membrane aeration was introduced and developed: LEAPmbr by SUEZ (previously known as GE Power & Water) and Mempulse™ by Evoqua (previously known as Siemens). In accordance to the data reported recently by municipal MBR plants (full-scale), the yearly average of specific energy consumption lies between 0.7–1.8, 0.8–1.1, 0.4–2.1, 0.4–0.6, 0.8–3.0, and 0.8–2.4 kWh/m3 in Germany, the Netherlands, Spain, China, Japan, and France, respectively. For optimized MBR (fully), aeration energy may lie in between 0.4–0.5kWh/m3, lower than 0.4kWh/m3 according to recent case studies.

7.5.4 Thickening of Sludge, Reduction, and Dewatering

Over the process of CAS, MBR has an advantage that its system is capable of minimizing sludge production; this is achieved by operating it at long SRT and a high concentration of sludge. While, in the process of sewage treatment, the production of sludge is unavoidable. From treatment processes of wastewater, a huge volume of sludge is dropped out and disposed of: out of total expense for wastewater treatment, around 60% is utilized on treating this sludge only. Furthermore, secondary pollution problems may be caused by using conventional disposal methods of ocean dumping and landfilling. Hence, methods which may reduce the excessive sludge are of more importance. Till now, numerous schemes have been started to diminish sludge production, involve uncoupling metabolism, mechanical solubilization, lysis-cryptic growth, heat treatment, bacteria predation, and maintenance metabolism. For reducing the volume of excessive sludge, all the above-discussed techniques can be combined into processes including final waste lines, sludge, and wastewater (Prado et al., 2017).

7.5.5 Wastewater Reuse and Reclamation by Aerobic MBRs

Increasing water shortage at a global level has increased the awareness and now actions are being taken for wastewater reuse and reclamation. Recycling of wastewater has numerous advantages of being environmentally friendly, cost-saving, and dependable. In due course, the increasing focus on the recycling of water has assisted research activities for the validation and development of high-quality and reliable treatment processes at relatively low expense. As MBR has the finest treated effluent, it is regarded as a superb tool for the reclamation of wastewater (López-Fernández et al., 2012; Nguyen et al., 2013). Nonetheless, conventionally used MBR systems developed with UF or MF membrane modules are not capable of removing trace organic contaminants and some pathogens. Hence, treated wastewater might be reutilized for only those procedures that require low quality of water like toilet flushing, cooling, and cleaning. For the purposes that require better quality of water, other techniques such as osmotic membrane filtration and membrane distillation can be combined into the MBR process (Besha et al., 2017; Bernal et al., 2018).

REFERENCES

1. Abass, O. K., Wu, X., Guo, Y., & Zhang, K. (2015). Membrane bioreactor in China: a critical review. *Int. J. Membr. Sci. Technol.*, *2*(2), 29-47.
2. Achilli A, Marchand EA, & Childress AE (2011). A performance evaluation of three membrane bioreactor systems: aerobic, anaerobic, and attachedgrowth. Water Sci Technol 63, 2999–3005
3. Ang, W. L., & Mohammad, A. W. (2019). Integrated and hybrid process technology. In *Sustainable Water and Wastewater Processing* (Vol.1,pp. 279-328). Elsevier.
4. Bernal, M. Á., Boluda Botella, N., & Prats, D. (2018). Mechanisms Involved in the Removal of Emerging Contaminants by a Combined UASB-MBR System., 1, 1-44
5. Besha, A. T., Gebreyohannes, A. Y., Tufa, R. A., Bekele, D. N., Curcio, E., & Giorno, L. (2017). Removal of emerging micropollutants by activated sludge process and membrane bioreactors and the effects of micropollutants on membrane fouling: A review. *Journal of environmental chemical engineering*, *5*(3), 2395-2414.
6. Bouhabila EH, Ben-Aim R, & Buisson H (2001). Fouling characterisation in membrane bioreactors. Sep Purif Technol 22–23, 123–132
7. Cote, P., Alam, Z., & Penny, J. (2012). Hollow fiber membrane life in membrane bioreactors (MBR). *Desalination*, *288*, 145-151.
8. Daigger, G. T. (2014). Ardern and Lockett remembrance. *Activated Sludge-100 Years and Counting*, 1, 1-15.
9. Dolar, D., Gros, M., Rodriguez-Mozaz, S., Moreno, J., Comas, J., Rodriguez-Roda, I., & Barceló, D. (2012). Removal of emerging contaminants from municipal wastewater with an integrated membrane system, MBR–RO. *Journal of hazardous materials*, *239*, 64-69.
10. Fane AG, & Fane SA (2005). The role of membrane technology in sustainable decentralized wastewater systems. Water Sci Technol 51(10), 317–325
11. Fawehinmi F, Lens P, Stephenson T, Rogalla F, & Jefferson B (2005). The influence of operating conditions on extracellular polymeric substances (EPS), soluble microbial products (SMP) and bio-fouling in anaerobic membrane bioreactors, 1,1-55.
12. Gómez-Espinosa, R. M., & Arizmendi-Cotero, D. (2017).

Role of membrane on emerging contaminant removal. In *Ecopharmacovigilance* (Vol.1, pp. 157-174). Springer, Cham.

13. Holloway, R. W., Achilli, A., & Cath, T. Y. (2015). The osmotic membrane bioreactor: a critical review. *Environmental Science: Water Research & Technology*, *1*(5), 581-605.
14. Hong SP, Bae TH, Tak TM, Hong S, & Randall A (2002). Fouling control in activated sludge submerged hollow fiber membrane bioreactors. Desalination 143, 219–228
15. Ioannou-Ttofa, L., Foteinis, S., Chatzisymeon, E., & Fatta-Kassinos, D. (2016). The environmental footprint of a membrane bioreactor treatment process through life cycle analysis. *Science of the Total Environment*, *568*, 306-318.
16. Jenkins, D., Richard, M. G., & Daigger, G. T. (2003). *Manual on the causes and control of activated sludge bulking, foaming, and other solids separation problems (Vol.1, pp. 1-43)*. Crc Press.
17. Jeon, S., Rajabzadeh, S., Okamura, R., Ishigami, T., Hasegawa, S., Kato, N., & Matsuyama, H. (2016). The effect of membrane material and surface pore size on the fouling properties of submerged membranes. *Water*, *8*(12), 602.
18. Jhaveri, J. H., & Murthy, Z. V. P. (2016). A comprehensive review on anti-fouling nanocomposite membranes for pressure driven membrane separation processes. *Desalination*, *379*, 137-154.
19. Judd S (2005). Fouling control in submerged membrane bioreactors. Water Sci Technol 51(6–7), 27–34
20. Judd S (2006). The MBR book: principles and applications of membrane bioreactors in water and wastewater treatment (Vol.1, pp. 1-44). Elsevier Science, Oxford
21. Judd, S. (2010). *The MBR book: principles and applications of membrane bioreactors for water and wastewater treatment(Vol.1, pp. 1-45)*. Elsevier.
22. Kang Y-T, Cho Y-H, & Chung E-H (2007). Development of the wastewater reclamation and reusing system with a submerged membrane bioreactor combinedbio-filtration. Desalination 202, 68–76
23. Krzeminski, P., Leverette, L., Malamis, S., & Katsou, E. (2017). Membrane bioreactors–a review on recent developments in energy reduction, fouling control, novel configurations, LCA and market prospects. *Journal of Membrane Science*, *527*, 207-227.

24. Lawrence P, Adham S, Barro L (2002). Ensuring water re-use projects succeed – institutional and technical issues for treated wastewater re-use. Desalination 152, 291–298
25. Le-Clech P, Chen V, & Fane TAG (2006). Fouling in membrane bioreactors used in wastewater treatment. J Membr Sci 284, 17–53
26. Le-Clech P, Fane A, Leslie G, & Childress A (2005). MBR focus: the operators' perspective. Filtr Sep 42(5), 20–23
27. Li, C., Cabassud, C., & Guigui, C. (2015). Evaluation of membrane bioreactor on removal of pharmaceutical micropollutants: a review. *Desalination and water treatment*, *55*(4), 845-858.
28. López-Fernández, R., Martínez, L., & Villaverde, S. (2012). Membrane bioreactor for the treatment of pharmaceutical wastewater containing corticosteroids. *Desalination, 300*, 19-23.
29. Luo, W., Phan, H. V., Li, G., Hai, F. I., Price, W. E., Elimelech, M., & Nghiem, L. D. (2017). An osmotic membrane bioreactor–membrane distillation system for simultaneous wastewater reuse and seawater desalination: performance and implications. *Environmental science & technology*, *51*(24), 14311-14320.
30. Mailler, R., Gasperi, J., Rocher, V., & Chebbo, G. (2013, June). Removal of priority and emerging substances by biological and tertiary treatments, 1, 1-33.
31. Mateo-Sagasta, J., Raschid-Sally, L., & Thebo, A. (2015). Global wastewater and sludge production, treatment and use. In *Wastewater* (Vol.1, pp. 15-38). Springer, Dordrecht.
32. Meng FG, Chae SR, Drews A, Kraume M, Shin HS, & Yang FL (2009). Recent advances in membrane bioreactors (MBRs): membrane fouling and membrane material. Water Res 43(6), 1489–1512
33. Meng, F., Chae, S. R., Shin, H. S., Yang, F., & Zhou, Z. (2012). Recent advances in membrane bioreactors: configuration development, pollutant elimination, and sludge reduction. *Environmental engineering science*, *29*(3), 139-160.
34. Mert, B. K., Ozengin, N., Dogan, E. C., & Aydıner, C. (2018). Efficient Removal Approach of Micropollutants in Wastewater Using Membrane Bioreactor. *Wastewater and Water Quality*, 41, 1-55.
35. Morrow, C. P., Furtaw, N. M., Murphy, J. R., Achilli, A., Marchand, E. A., Hiibel, S. R., & Childress, A. E. (2018). Integrating an aerobic/anoxic osmotic membrane bioreactor with membrane distillation for

potable reuse. *Desalination, 432*, 46-54.

36. Nguyen, L. N., Hai, F. I., Kang, J., Price, W. E., & Nghiem, L. D. (2013). Removal of emerging trace organic contaminants by MBR-based hybrid treatment processes. *International Biodeterioration & Biodegradation, 85*, 474-482.
37. Ong, S. L., & Hu, J. (2008). Removal of Emerging Contaminants in Water—Do We Have Solutions?. *Water Environment Research*, 1, 771-773.
38. Phattaranawik, J., Fane, A. G., Pasquier, A. C., & Bing, W. (2008). A novel membrane bioreactor based on membrane distillation. *Desalination, 223*(1-3), 386-395.
39. Prado, M., Borea, L., Cesaro, A., Liu, H., Naddeo, V., Belgiorno, V., & Ballesteros Jr, F. (2017). Removal of emerging contaminant and fouling control in membrane bioreactors by combined ozonation and sonolysis. *International Biodeterioration & Biodegradation, 119*, 577-586.
40. Psoch C, & Schiewer S (2005). Critical flux aspect of air sparging and backflushing on membrane bioreactors. Desalination 175, 61–71
41. Psoch C, & Schiewer S (2006). Resistance analysis for enhanced wastewater membrane filtration. J Membr Sci 280, 284–297
42. Sauvé, S., & Desrosiers, M. (2014). A review of what is an emerging contaminant. *Chemistry Central Journal, 8*(1), 1-7.
43. Shannon, M. A., Bohn, P. W., Elimelech, M., Georgiadis, J. G., Marinas, B. J., & Mayes, A. M. (2010). Science and technology for water purification in the coming decades. In *Nanoscience and technology: a collection of reviews from nature Journals* (Vol.1, pp. 337-346).
44. Smith PJ, Vigneswaran S, Ngo HH, Ben-Aim R, & Nguyen H (2005). Design of a generic control system for optimising back flush durations in a submerged membrane hybrid reactor. J Membr Sci 255:99–106
45. Stephenson T, Judd S, Jefferson B, Brindle K (2000) Membrane bioreactors for wastewater treatment (Vol.1, pp. 1-33). IWA Publishing, London
46. Tajima, F., & Yamamoto, T. (1988). *U.S. Patent No. 4,756,875,, pp. 1-30*. Washington, DC: U.S. Patent and Trademark Office.
47. Tangsubkul, N., Parameshwaran, K., Lundie, S., Fane, A. G., & Waite, T. D. (2006). Environmental life cycle assessment of the microfiltration process. *Journal of membrane science, 284*(1-2), 214-226.

48. Verrecht, B., Maere, T., Nopens, I., Brepols, C., & Judd, S. (2010). The cost of a large-scale hollow fibre MBR. *Water Research*, *44*(18), 5274-5283.

49. Wang ZW, & Wu ZC (2009). A review of membrane fouling in MBRs: characteristics and role of sludge cake formed on membrane surfaces. Sep Sci Technol 44(15), 3571–3596

50. Wang, J., Pathak, N., Chekli, L., Phuntsho, S., Kim, Y., Li, D., & Shon, H. K. (2017). Performance of a novel fertilizer-drawn forward osmosis aerobic membrane bioreactor (FDFO-MBR): Mitigating salinity build-up by integrating microfiltration. *Water*, *9*(1), 21.

51. Wang, X., Chang, V. W., & Tang, C. Y. (2016). Osmotic membrane bioreactor (OMBR) technology for wastewater treatment and reclamation: Advances, challenges, and prospects for the future. *Journal of membrane science*, *504*, 113-132.

52. Wang, Y. K., Sheng, G. P., Li, W. W., Huang, Y. X., Yu, Y. Y., Zeng, R. J., & Yu, H. Q. (2011). Development of a novel bioelectrochemical membrane reactor for wastewater treatment. *Environmental science & technology*, *45*(21), 9256-9261.

53. Wanner, J. (1994). *Activated sludge: bulking and foaming control (Vol.1, pp. 1-33)*. CRC Press.

54. Xiao, K., Liang, S., Wang, X., Chen, C., & Huang, X. (2019). Current state and challenges of full-scale membrane bioreactor applications: a critical review. *Bioresource technology*, *271*, 473-481.

55. Yang W, & Cicek N, Ilg J (2006). State of the art of membrane bioreactors: worldwide research and commercial applications in north America. J Membr Sci, 270, 201–211

56. Zgavarogea, R., Covaliu, C., Niculescu, V., Miricioiu, M., Paun, N., Iordache, A., & Neacşa, M. (2016). MEMBRANE BIOREACTOR (MR) TECHNOLOGY-A NEW AND EFFECTIVE TREATMENT OF WASTEWATER. *Progress of Cryogenics & Isotopes Separation*, *19*(2), 1-33.

57. Zhang, J., Chua, H. C., Zhou, J., & Fane, A. G. (2006). Factors affecting the membrane performance in submerged membrane bioreactors. *Journal of membrane science*, *284*(1-2), 54-66.

58. Zhang, M., Lee, E., Vonghia, E., Hong, Y., & Liao, B. (2020). Introduction to aerobic membrane bioreactors: Current status and recent developments. In *Current Developments in Biotechnology and Bioengineering* (Vol.1, pp. 1-23). Elsevier.

CHAPTER

8

WATER TREATMENT USING ANAEROBIC MEMBRANE BIOREACTORS

CONTENTS

8.1 INTRODUCTION

AnMBRs(Anaerobic membrane bioreactors) are being used gradually in industrial wastewater treatment as the technology signifies a cost-effective substitute to that based on aerobic processes. AnMBRs (Anaerobic membrane bioreactors)are highly efficient not only in decreasing chemical oxygen demand as well as in the organic matter removed is transformed into a suitable energy source—biogas. AnMBRs (Anaerobic membrane bioreactors) while occupying a small footprint produces an effluent that is rich in nutrients and free of pathogens and solids. AnMBRs improve performance while dealing with toxic or inhibitory substrates as the membrane holds biomass, typical of industrial wastewaters (Liao et al., 2006; Lin et al., 2013). However, some problems remain, including poor efficiency at a lower temperature (Anaerobic membrane bioreactors are usually operated at thermophilic or mesophilic conditions)as well as membrane fouling and its related effects. More research is needed on the removal of precise micro-pollutants, the inclusion of efficient dissolved methane recovery, lowering hydraulic retention time, establishing quantitative mass and energy/economic balances, and exclusion of nutrients. The applications, limitations, and perceptions of AnMBRs are evaluated and summarized in this chapter with the importance of industrial wastewater treatment. TheAnaerobic membrane bioreactors furthermore are related to other currently available wastewater treatment technologies (Ozgun et al., 2013; Dvořák et al., 2014).

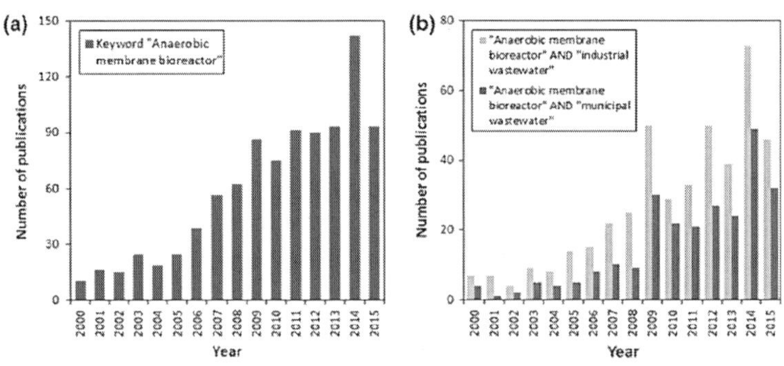

Figure 8.1: The number of publications over the past 15 years (Scopus) using the key phrases; (a) "AnMBR" only, and (b) "AnMBR" in combination with "industrial wastewater" and "municipal wastewater".

[Source: https://www.tandfonline.com/doi/full/10.1080/19443994.2015.1100879]

High effluent quality is attained by combining with membrane separation, with no total bacteria or solids in the effluent. By combining the advantages of anaerobic treatment with membrane separation,AnMBRs (Anaerobic membrane bioreactors), makes the treatment of a broad spectrum of wastewaters an attractive approach (Dereli et al., 2012; Ramos et al., 2014). Anaerobic membrane bioreactors are useful for treating wastewaters because of their high operation constancy under extreme conditions, including high salinity, poor biomass granulation, or high suspended solids content. Thus, Anaerobic membrane bioreactors are now of great interest to both industrial and the researcher's community, with many pilot-scale - and laboratory studies taking on, particularly as regards the extensive range of industrial wastewaters needing treatment (Le-Clech, 2010; Diez et al., 2012). The notable rise in interest in AnMBR is imitated in the number of peer-reviewed publications associated to industrial/municipal wastewater treatment and AnMBR more than past 15 years found in Scopus (Fig. 8.1(a)—keyword "AnMBR", Fig. 8.1(b)—"AnMBR" plus "industrial wastewater", and "municipal wastewater"). AnMBR configurations are shown in Figure 8.2 below.

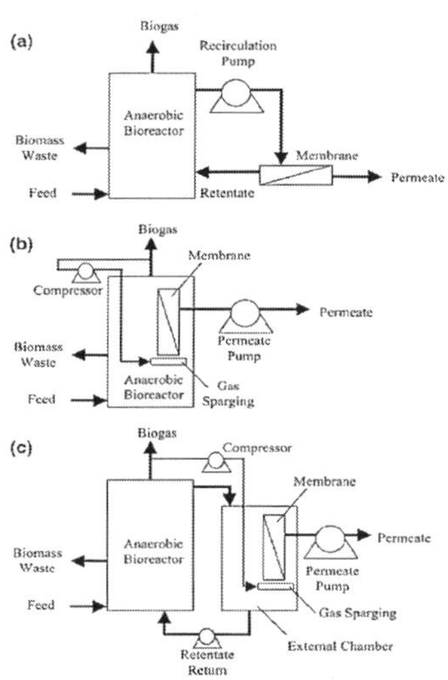

Figure 8.2: AnMBR configurations: (a) side-stream (external, cross-flow) configuration—pressure-driven membrane, (b) submerged configuration with membrane immersed directly in the reactor—vacuum-driven membrane, and (c) submerged configuration with the membrane in the separate chamber—vacuum-driven membrane.

[Source: https://www.tandfonline.com/doi/full/10.1080/19443994.2015.1100879]

As regards industrial wastewater treatment, we present here a mini-review, based on a complete literature survey of the current perspectives, limitations, and applications for anaerobic membrane bioreactors. Furthermore, including membrane materials and system configurations, we present a summary of the Anaerobic membrane bioreactor and discuss the disadvantages and advantages in comparison with the currently available other wastewater treatment technologies. Lastly, we address the needs for further research and future perspectives.

8.2 FUNDAMENTALS OF ANMBR

8.2.1 Reactor Configuration

The membrane used in Anaerobic membrane bioreactors can be located in either of two places in AnMBR (Anaerobic membrane bioreactor) system as the membrane used in AnMBRs turns as a total barrier to micro-organisms, it resulting in two main AnMBR configurations: pressure-driven cross-flow with the exterior membrane (side-stream; Fig. 8.2(a)) and vacuum-driven, with the membrane immersed directly in the anaerobic reactor (Fig. 8.2(b)). An additional configuration combines the external placement of the membrane, which is submerged in a vacuum-driven operation and a separate chamber (Fig. 8.2(c)).

8.2.1.1 Side-stream Configuration

The recirculation pump safeguards essential trans-membrane pressure in the membrane chamber, in the side-stream configuration, (Fig. 8.2(a)). So, the formation of a filtration cake on the membrane's surface can permanently be disrupted by the crossflow velocity. As the side-stream configuration process consumes energy, any residual energy can be used to mix the suspension in the anaerobic reactor. To attain optimum operational efficiency about energy demand and filtration cake removal the cross-flow velocity along the membrane's surface is usually kept inside a range of 2–4 m s−1. (Dvořák et al., 2011; Bornare et al., 2014). The breakdown of larger sludge flocs and agglomerates is of importance here because of high shear forces, which guides to a reduction in biomass particle size. The more rapid fouling and membrane clogging can be resulted by the subsequent release of soluble microbiological products (Lesjean et al., 2005; Meng et al., 2009). The clogging of membrane pores indeed is a serious and well-known issue related to membrane fouling (Zsirai et al., 2012). The chemical cleaning that requires by the fouled membranes and this can easily be accomplished by using the side stream configuration because the membrane can be cleaned without interrupting system operation and is more accessible, different than the submerged membranes that are engrossed directly into the anaerobic reactor. Due to some reason, membrane replacement or inspection is also much easier (Cote & Thompson, 2000; Andrade et al., 2014).

8.2.1.2 Submerged Membrane

The advantage of submerging the membrane directly in the anaerobic reactor lies in the absence of a recirculation pump in the membrane compartment (Fig. 8.2(b)). Thus, energy consumption is lower than in the side-stream configuration. Furthermore, the absence of a recirculation pump means that there lowered the shear forces and there is no cross-flow effect, which results in less micro-organism stress (Aslan et al., 2014). By using the external stirrer or the produced biogas disruption of filtration cake formation on the membrane's surface, along with anaerobic reactor mixing is typically ensured. However, mixing with biogas can guide to functioning problems, especially when contaminated compounds occur in the feed or during AnMBR start-up. The activity of the anaerobic groups can be inhibited, in these circumstances decreasing biogas production to levels very low to safeguard suitable mixing in the anaerobic reactor. To avoid such a situation a mechanical stirrer is usually installed into the reactor. The mixing with either a mechanical stirrer or biogas is less effective in eliminating filtration cake from the membrane's surface compared with the side-stream configuration, which means that membranes become contaminated quicker (Zhang et al., 2015). Chemical cleaning of submerged membranes is tough as AnMBR operation, as mentioned above, and membranes have to be removed before cleaning and have to be interrupted. To achieve the same hydraulic performance as attained with the side-stream configuration, a larger membrane surface area with submerged membranes is needed (Santos & Judd, 2010; Pandey & Singh, 2014).

8.2.2 Membrane Materials and Modules used in AnMBR

The membrane materials, in general, can be divided into three main categories: metallic, ceramic and polymer. Compared to metallic or ceramic membranes polymer membranes have the advantage of lesser cost; hence for a broad range of different applications, they are preferred, with metallic or ceramic membranes being used for particular applications. In AnMBRs the polymer membranes used are usually based on polysulfone (PSF), polypropylene (PP), polyethylene (PE) polyethersulfone (PES), orpolyvinylidene difluoride (PVDF) (Vyrides & Stuckey, 2009; Jeong et al., 2010). Polymer membranes also have some drawbacks, though, including reduced stability and lower permeability toward chemical cleaning compared with metallic or ceramic membranes (Judd, 2010; Stuckey, 2012).

Ceramic membranes are a more appropriate option for AnMBRs than polymer membranes from an operative point of view. Membrane fouling under anaerobic conditions has a tendency to be more pronounced than in aerobic MBRs (Deowan et al., 2015); meaning that higher temperatures and/or longer exposure, more concentrated chemicals is vital to recovering membrane permeability. Since ceramic membranes have increased resistance to abrasion and corrosion, as well as outstanding chemical thermal and stability, they can be cleaned more effectually as compared to polymer membranes (Ersu & Ong, 2008).

While metallic membranes show greater tolerance to high temperatures and oxidation, easier permeability recovery after fouling, enhanced stability, higher hydraulic performance than polymer membranes, but they are expensive, hence used only in precise applications.

Most AnMBRs now use ultrafiltration or PVDF microfiltration membranes according to the conducted literature survey, (Table 8.1). A flat-sheet dynamic membrane Only Xie et al. (2014) have experienced in AnMBR treating the landfill leachate. Unlike the ultrafiltration or microfiltration membranes, the performance and characteristics of dynamic membranes are determined particularly by concentration, molecular weight, shape, type of solution being filtered along with hydrodynamic conditions along the membrane.

In present AnMBRs hollow-fiber membrane modules tend to be the most commonly used (Table 8.1) likely because of their cost efficiency and high packing density, subsequently, flat-sheet membrane modules, which display great stability and are easily replaced or cleaned when defective (Kanai et al., 2010).

Despite having a low fouling tendency, because of being easy to replace or clean, and high cross-flow velocities, tubular membrane modules are hardly used nowadays as having high dead volume and low packing density (Dereli et al., 2014; Wallace & Safferman, 2014). Table 8.1 shows full summary about membrane characteristics used in present AnMBRs

8.3 COMPARISON WITH OTHER TECHNOLOGIES AND ADVANTAGES OF ANMBR

Many industrial wastewaters comprise great concentrations of biodegradable organic matter, making them appropriate as feed for anaerobic treatment (Wang et al., 2013; Jensen et al., 2015). Some industries usually used anaerobic technologies like expanded granular sludge beds, hybrid UASBs, up-flow anaerobic sludge beds (UASBs), anaerobic filters, or interior circulation reactors can accomplish high organic matter elimination efficiencies with such wastewaters, often reaching more than 90%. The organic matter is transformed into biogas at the same time. However, wastes from these reactors, often contain suspended solids and organic matter. For instance, speech recorded suspended solids at greater than 1 g L−1 in the effluent from a UASB (Speece, 1983; Choi et al., 2013). The biomass characteristics (mostly granulation properties) moreover, can easily be affected in these systems, with biomass afterward washed out through effective problems, e.g. when toxic compounds arise in the treated wastewater or through fast system overloading. Because of the membrane, AnMBRs are capable of hydraulic peaks and fade concentration unlike common anaerobic technologies and hence tolerate fluctuations in organic loading. The membrane safeguards that biomass is parted from the effluent; therefore, this technology displays great potential for the treatment of those wastewaters that harmfully influence granular biomass in high-rate anaerobic reactors (Blaszczyk et al., 1994; Couras et al., 2014).

Table 8.1: Membrane characteristics, including permeabilities (calculated based on presented data) in recent laboratory-scale (L), pilot-plant scale (P) and full-scale (F) AnMBRs; (De Vrieze et al., 2014; Kijjanapanich et al., 2014; Ng et al.,2014)

Scale	Set-up	Wastewater source	Permeability (LHM bar^{-1})	Mem-brane	Module	Material	Supplier
P	In.	Slaughter-house (side A)	625–50	UF	HF; ZW-10	PVDF	Zenon
La	In.	Molasses	92	UF	FS	Chlori-nated PE	Kubota
P	In.	Slaughter-house (side B)	40–8	UF	HF; ZW-10	PVDF	Zenon
L	Ex.	Bamboo industry	97–14	UF	HF	PVDF	–

Lb	In.	Molasses	96	UF	FS	Chlorinated PE	Kubota
L	In.	Landfill leachate	1,200–15	Dynamic	–	–	–
Pd	Ex.	Liquid dairy manure	10	UF	Tubular	PVDF	X-flow
L	Ex.	Pharmaceutical industry	400–13	UF	HF	PVDF	GE-LS1
L	Ex.	Lipid rich corn-to ethanol thin stillage	–	UF	Tubular	PVDF	Pentair X-Flow
Pc	Ex.	Liquid dairy manure	62	UF	Tubular	PVDF	X-flow
L	Ex.	Debris leachate	7–2	UF	HF	PVDF	Norit-X flow
P	In.	Food industry (snacks)	30–7	UF	HF	PVDF	Porous Fibre Spain
Le	Ex.	Pharmaceutical industry	500–13	UF	HF	PVDF	GE-LS1
L	In.	Food industry	200–114	MF	FS	Chlorinated PE	Kubota
F	In.	Food industry (salad dressings)	417–42	MF	FS	–	Kubota
L	In.	Food industry (sugarcane vinasse)	50–10	MF	HF	Polyetherimide	Pam Membranes Seletivas
P	In.	Municipal wastewater	40	UF	FS	PES	Microdyn-Nadir

Note:

UF—ultrafiltration;

In.—internal (submerged) set-up;

En.—external (side-stream) set-up,

FS—flat sheet

HF—hollow-fiber,

MF—microfiltration,

Several studies have already confirmed stable biogas production, high treatment effectiveness, and high operational stability in AnMBRs under extreme conditions. Along with the separation of suspended solids, AnMBRs are also capable of retaining pathogens and bacteria from effluent and at lower energy consumption levels (Chan et al.,2009; Al-Hashimia et al., 2013). As the effluent from AnMBR comprises nutrients, like phosphorus and nitrogen, and shows a greater COD than effluents from aerobic MBRs, it can be recycled for non-potable purposes, particularly for process waters or irrigation (depending on desirable final water quality). To fulfill local standard discharge limits when effluent is discharged into water bodies, its posttreatment can be needed (Ellouze et al., 2009; Ho & Sung, 2010).

AnMBR effluent quality tends not to be affected by changes in biomass settling or granulation properties, unlike that from other anaerobic treatment technologies. Finally, fast system operation recovery is achieved due to the complete retention of biomass by the membrane. According to Tao et al. (2012) and van Lier (2008), the membrane represents a total barrier to the slow-growing micro-organisms that can remove specific pollutants from industrial wastewaters as they accumulate in the system, regardless of hydraulic retention time (HRT). Tao et al., for example, increased the activity of slow-growing Anammox micro-organisms by 19 times through their retention by a membrane (Tao et al., 2012).

AnMBRs give other important operational advantages along with those related to the membrane. For instance, total energy consumption is decreased as no oxygen is needed for biotransformation of organic matter. Furthermore, the treatment of organic matter yields biogas as a suitable end product. A renewable energy source i.e. Biogas is typically combusted to produce heat and electricity. For anaerobic digestion processes, the heat is then used to ensure suitable temperatures in the reactor. Note, that most studies however to date have been undertaken under mesophilic (35–37°C) conditions and there is a common deficiency of studies conducted underneath thermophilic temperature (only study by Qiao et al. (2013), regardless of enhanced filtration because of better slush rheological properties and the option to work under higher organic loading rates under these circumstances (Meabe et al., 2013; Qiao et al., 2013). This deficiency of studies with thermophilic temperature is possible because of the danger of deterioration in performance-related with the buildup of volatile fatty acids, a follow-on in the inhibition of microbial activity. Meabe et al. (2013) noted enlarged membrane fouling and only to some extent higher biogas production under thermophilic conditions related to mesophilic temperature.

Despite the fact significantly lower functioning costs are predictable under psychrophilic conditions, only an inadequate number of studies have been led under these circumstances. Martinez-Sosa et al. also worked an AnMBR pilot-plant at 20°C; though, the functioning period lasted for just 21 d, as a result of which the temperature was raised to 35°C. Furthermore, the public wastewaters treated in this pilot-plant AnMBR were differentfrom those in the earlier mentioned studies (Skouteris et al., 2012).

As no oxygen is required operational costs in AnMBR plants are expressively reduced, and a huge fraction of the heating and electricity required to function the plant can be covered by the biogas produced. On biomass production, the degree to which these costs are covered will depend. For example, total AnMBR costs for treatment of Kraft mill effluent, were expressively lesser than those by using aerobic treatment (Lin et al., 2011; Wijekoon et al., 2012). Thus, the largest percentage of both capital and operational costs during AnMBR operation is attributable to the membrane itself and aspects associated with membrane fouling. Membrane foulingis considered the main disadvantage which reduces flux avoiding widespread application and quicker commercialization of membrane technology in wastewater treatment (Minami, 1994; Aquino et al., 2006). Membrane fouling decreases membrane lifetime and flux, together with gas scouring or cross-flow pumping of the membrane, all have to be considered into the total operative costs directly related to the use of the membrane. In part due to this, AnMBRs are generally worked at lower fluxes than aerobic MBRs (Dereli et al., 2015).

In AnMBR excess sludge production is lesser than in aerobic MBRs; associated with aerobic MBRs up to 20 times, thus further reducing operational costs. Finally, AnMBRs also have the benefit of a shorter start-up period related to UASB reactors, while the start-up period for UASB reactors can take numerous months, both Stuckey and Hu (2006) and Lin et al. (2011) have described start-up times of just 6 and 12 d, respectively, for AnMBRs.

When related with the further technologies used in wastewater treatment, e.g. biofilm filtration, advanced oxidation processes (AOPs) and activated sludge process, AnMBRs tend to be not as much of expensive than AOPs and more effective than either biofilm/biofiltration or activated sludge systems when treating industrial wastewaters underneath extreme conditions, e.g. presence of fat and oil, high salinity, high suspended solids content or other inhibiting compounds (Lew et al., 2009).

8.4 DISADVANTAGES AND LIMITATIONS OF ANMBR

Whereas AnMBRs exhibit many advantages over conventional systems, many problems remain. As in the aerobic MBRs, the most critical disadvantage in AnMBRs is the fouling of the membrane. Fouling, which outcomes in reduced hydraulic performance, has restricted the extensive application of the membrane technology. Membrane fouling is an intricate issue that is affected by the range of factors, comprising influent characteristics, operational conditions, biomass and membrane properties, and their common combination (Seghezzo et al., 1998; Álvarez et al., 2006). As an outcome, an extensive range of problems associated with the fouling of membrane have been the topic of intensive study. Membrane fouling is primarily caused by accumulation and deposition of micro-organisms, solutes, colloids, and cell debris inside or on the membrane. Precipitation of the inorganic compounds, primarily struvite (MgNH4PO4; magnesium ammonium phosphate), has also been recognized as a significant part of the irreversible fouling on membranes in the AnMBRs. Other inorganic salts like calcium carbonate (CaCO3) and potassium struvite (K2NH4PO4) have been also recognized in the fouling layer (Van Lier, 2008). Both operational parameters and membrane properties can play a vital part in the rate at which the inorganic compounds are generally precipitated. Meabe et al. (2013), for instance, reported that fouling by struvite amplified at higher operating temperatures (55 degrees C, compared to 35 degrees C) because of augmented ammonia nitrogen concentrations (Yu et al.,2013).

8.5 OPERATIONAL EXPENSES OF ANMBR

Costs linked with the membrane are a vital issue restricting the extensive application of AnMBR. Even though membrane costs have reduced considerably over recent years, the capital expenses of the membrane and mainly operational costs linked with filtration procedure still signify one of the major disadvantages of the membrane bioreactors. For instance, according to Pretel et al. (2014) up to 85 to 90 percent of the power necessities in AnMBR is associated with the filtration process and the membrane fouling mitigation. Most of the energy is used up by membrane tank biogas recycling blower or the membrane tank sludge feeding pump demonstrating up to 75 percent of the total demand of energy in AnMBR (Meng et al., 2007; Boyle-Gotla et al., 2014).

Generally, there have been quite a few studies that have estimated the costs linked with operating AnMBRs, predominantly with those that treat the industrial wastewaters. One example, Ferrer et al. (2015) described operational and capital expenditure for AnMBR treating both low-sulfate and sulfate-rich municipal wastewaters. These authors described a clear dependency among increasing MLSS (mixed liquor suspended solids) concentration and the membrane scouring, utilization of chemical reagents and the operational costs. The optimal MLSS concentration for the AnMBR was found at 12 g L−1, with the linked specific demand of energy of around 0.22 kW h m−3 of treated wastewater during the operation with no energy recovery. When the biogas was re-utilized, the final specific demand of energy dropped to 0.14 kW h m−3. Similarly, Pretel et al. (2014) described the energy necessities of AnMBR treating the sulfate-rich urban wastewaters. They described the specific demand of energy in the range of 0.11 to 0.49 kW h m−3 when the biogas was captured. Improvement in the energy balance of AnMBR might be also accomplished through seizing the biogas methane and methane dissolved in the waste or functioning AnMBR at high surrounding temperature or high SRT (solids retention times).

Compared to the aerobic MBR, for example, Gabarro´n et al. (2014) stated values for total specific demand of energy in the full-scale aerobic MBR reaching the value of 1.54 kW h m−3 for the flat-sheet membrane and value 1.12 kW h m−3 for the hollow fiber membrane. Even after the execution of their energy-saving approaches involving primarily optimization of both membrane air-scouring and biological aeration, the specific demand of energy reached the values of 1.12 kW h m−3 and 0.71 kW h m−3, respectively, irrespective of similar annually-averaged hydraulic loads (Wei et al., 2014).

Martin et al. (2011) assessed both the anaerobic and aerobic MBRs and reported the total specific demand of energy around 2 kW h m−3 for the aerobic MBR with entire sludge retention, whereas in AnMBR the demand of energy varied from 0.03-5.7 kW h m−3. The highest demand for energy was perceived as an outcome of increased demand for gas for extreme membrane fouling control (Araya et al., 1999).

Another instance stating AnMBR operational expenses is studied by Smith et al. (2014). These authors related, based on the outcomes of process modeling and the system analyses, AnMBR with the other technologies involving conventional activated sludge and aerobic MBR, high-rate activated sludge, all coupled with the anaerobic digestions. It was concluded that AnMBR can recuperate more energy for high and medium strength domestic wastewaters as compared with the evaluated technologies. The energy recovery is strongly dependent on the temperature and composition of the wastewater, pre-treatment processes, or operational conditions (Ivanovic & Leiknes, 2012).

As mentioned previously, permeability's in AnMBRs is generally lower as compared to those in aerobic MBRs. whereas permeabilities in the aerobic MBRs are usually around 200 LHM bar−1, as can be noticed from Table 8.1, the usual permeabilities were normally within the range from ten to 150 LHM bar−1 in recent AnMBRs. The highest primary permeability reached 1,200 LHM bar−1, conversely, it was noticed for the dynamic membrane that is generally characterized by high hydraulic performance. Also, the preliminary permeability of 500 and 400 LHM bar−1 for the ultrafiltration membrane stated by Ng et al. was caused by considerably lower MLSS concentration reaching around 7.2 and 1.7 g L−1, correspondingly, when compared with some other reported studies (Table 8.2). Lower hydraulic performance of the AnMBR together with the fouling of membrane has to be, thus, considered when computing the total cost balance (Dos Santos et al., 2007).

Table 8.2: A general idea of technological parameters utilized in current AnMBRs (Christian et al., 2011; Qiao et al., 2013)

Scale	Set-up	Wastewater source	Solids (g L^{-1})	HRT (d)	SRT (d)	Temp. (°C)
L	Ex.	Bamboo industry	8.0–14.0 (VSS)	2–10	–	28–30
L	In.	Molasses (concentrated)	18.7 (TSS)	26	1,535	34
L	In.	Molasses (diluted)	10.0 (TSS)	5.3	1,535	34
P	In.	Slaughterhouse (side A)	26.1 (TS)	2–7	50–1,000	37
P	In.	Slaughterhouse (side B)	17.0–40.2 (TS)	4–7	50	37
P	Ex.	Liquid dairy manure (phase 1)	54.0 (VS)	19 (10–35)	19 (10–35)	Room
P	Ex.	Liquid dairy manure (phase 2)	28.0 (VS)	12	24	Room
L	Ex.	Pharmaceutical industry	6.0–8.4 (MLSS)	21.3–42.6	700	27
L	Ex.	Debris leachate	6.0–7.0 (MLSS)	15.5	–	23
L	In.	Landfill leachate	16.9 (VSS)	2.5	125	37

P	In.	Food industry (oil and grease content)	11.4 (VSS)	–	–	36
La	Ex.	Pharmaceutical industry	1.2–2.1 + 11.0[b] (MLSS)	21.3–42.6	700	27
L	Ex.	Lipid rich corn-to-ethanol thin stillage	15.2–24.9 (VSS)	10.1	20; 30; 50	37
L	In.	Food industry	75.0 (MLSS)	20–70	20–100	57
F	In.	Food industry (salad dressings)	20.0–45.0 (MLSS)	–	–	33
P	In.	Food industry (snacks industry)	18.0 (VS)	–	–	30–36
L	In.	Food industry (sugar-cane vinasse)	20.0 (MLVSS)	–	–	19–27
P	In.	Municipal wastewater	15.0–20.0 (TSS)	–	–	35/20

Note: -*TSS—total suspended solids, TS—total solids, VSS—volatile suspended solids, VS—volatile solids, MLSS—mixed liquor suspended solids*

8.6 ANMBR APPLICATIONS IN THE INDUSTRIAL WASTEWATER TREATMENT

There had been few scientific studies centered on pilot-scale AnMBRs till 2012 (Saddoud et al, 2006), most of the studies being carried out under laboratory conditions having laboratory scale apparatus. Many laboratory-scale studies have been carried out on AnMBRs fed them with synthetic wastewater, which is normally used to authenticate the laboratory-scale performance of AnMBR or test new system/reactor concepts (Savant et al., 2006).

Till now, AnMBR pilot-plants have been utilized for the treatment of the different wastewaters having a high content of the organic matter, like that from industrial and food processing utilization, the paper and pulp industry, polymer synthesis and textile production. Most freshly, pilot and laboratory scale AnMBR plants have been utilized to treat diverse food processing wastewater, for example, from slaughterhouses, dairy manure, molasses production, landfill leachate, and pharmaceutical production. While several full-scale aerobic MBR types of research have been carried out on industrial or municipal wastewaters, to date there has been just studied stating on the full-scale AnMBR installation for the industrial wastewater treatment (Table 8.2). The full-scale AnMBR is engaged to treat wastewaters initiating from the production of salad dressings. The general idea about simple technological parameters of existing AnMBRs is briefed in Table 8.2 (Fenu et al., 2010; Gabarrón et al., 2014).

One of the chief operational benefits of an MBR is the probability of functioning the reactor at higher SRT as compared to the conventional digestion technology. Generally, high SRTs confirm high COD elimination efficiency. Moreover, high SRTs also aid micro-organisms to familiarize themselves with the diverse compounds existent in industrial wastewaters, several of which can be problematic to biodegrade, for example, those in saline wastewaters, pharmaceutical wastewaters, or both in an amalgamation. Following biomass adaptation, though, AnMBRs are proficient in accomplishing very high COD elimination efficiencies when fed with poorly biodegradable wastewater (Lin et al., 2009).

As can be viewed from Table 8.3, the lowest COD elimination efficiencies were discovered for liquid dairy dung, with the levels reaching around 40 percent. Note, conversely, that liquid dairy dung had a concentration of COD more than twice of the other feeds, with molasses wastewater and the lipid-rich corn to ethanol thin stillage with higher COD concentrations. Even then, nevertheless, elimination efficiencies reached 94.4 percent and >99 percent, correspondingly, in spite of AnMBR treating the lipid-rich corn to ethanol thin stillage being functioned under a comparable SRT and a lower HRT with that for handling liquid dairy dung. The AnMBR handling the molasses wastewater was functioned at a quite higher HRT (26 d) and a considerably higher SRT (1,535 d) as compared to the SRT functioning maximum of 50 d in AnMBR utilized for lipid-rich corn to ethanol thin stillage for the liquid dairy dung (Gao et al., 2010).

Table 8.3: Effluent and influent characteristics, comprising COD elimination efficiency, in recent AnMBRs (Martinez-Sosa et al., 2011; Mota et al., 2013)

Scale	Source of waste-water	Influent (g L^{-1})	Effluent (g L^{-1})	COD rem. (%)
L	Landfill leachate	COD 13.0; NH_4-N 3.2	COD 4.91	62
P	Slaughterhouse (side A)	COD 5.92; NH_4-N 0.024; TP 0.019	COD 0.07; NH_4-N 0.17; TP 0.014	95
P	Slaughterhouse (side B)	COD 10.6; NH_4-N 0.059; TP 0.036	COD 0.18; NH_4-N 0.29; TP 0.025	95

L	Bamboo industry	COD 21.4; NH_4-N 0.398	COD 1.5; NH_4-N 0.36	85–90
L	Molasses (concentrated)	COD 110.9; TKN 12.0	COD 10.7	94
L	Molasses (diluted)	COD 14.5; TKN 1.10	COD 0.5	93
P	Liquid dairy manure (phase 1)	COD 53.7	–	41
P	Liquid dairy manure (phase 2)	COD 41.8	–	42
La	Pharmaceutical industry	COD 15.4; TN 1.42; PO^3_4-P 0.17	COD 8.23	47
L	Pharmaceutical industry	COD 15.4; TN 1.42; PO^3_4-P 0.17	COD 8.77	43
L	Lipid rich corn-to-ethanol thin stillage	COD 72.2; $SO^2_4$0.95	COD 0.47	>99
L	Debris leachate	$SO^2_4$0.5–0.75	$SO^2_4$0.05–0.5, Ca^{2+} 0.2–0.5	80 (SO24)
F	Food industry (salad dressings)	COD 39.0; BOD_5 18.0	COD 0.21; BOD 0.02	99
P	Food industry (high oil and grease content)	COD 7.9–22.8; BOD_5 4.9–10.3; TKN 0.08–0.38; PO^3_4-P 0.06–0.16	COD 0.18–0.3	97
L	Food industry (sugarcane vinasse)	COD 17.7	COD 0.49	96
L	Food industry	–	COD 2.0–15.0	67
P	Municipal wastewater	COD 0.4	COD 0.08	90
P	Food industry (snacks industry)	COD 11.0; BOD_5 7.3; N 0.2, P 0.04	COD 1.7	75

Biological activity reduces with a drop in temperature and anaerobic systems, this outcome in a related decrease in the COD elimination efficiency. As an outcome, the majority of fresh studies have been conducted under mesophilic temperatures.

Thermophilic temperatures have rarely been applied (Table 8.2), with just Qiao et al. utilized this temperature for treatment of the food industry wastewater. Psychrophilic conditions, conversely, have often been applied in circumstances of low-strength domestic wastewater, wastewater, or under higher surrounding climate (Spagni et al., 2010).

8.7 THE PROSPECT OF ANMBRS AND ADDITIONAL RESEARCH NEEDED

It is normally accepted that the number of industrial wastewaters of intense composition will rise in the future. Consequently, upsurge in the application of efficient and more modern treatment technologies, like AnMBR is expected, particularly considering the present emphasis on sustainable production and lessening of carbon footprints. AnMBRs are especially significant as regards this last point because of their capability to yield energy in the biogas form. As an outcome, AnMBRs are possible to be applied more under the full-scale functioning conditions in the future and over a wider industrial spectrum (Wong et al., 2009).

Till now, AnMBRs have been intensively applied for industrial wastewater (Figure 8.1(b)), with only restricted full-scale applications for municipal wastewater treatment. Whereas this is perhaps due in part to comparative innovation of the technology, restrictions linked with fouling of the membrane are also a vital issue restricting more widespread application. The significant struggle must be put into fixing this issue, thus, before AnMBRs are more extensively accepted. Below, various possible methods of reducing fouling and improving membrane performance, and research necessities are discussed.

8.7.1 Operational Conditions

Many studies have been conducted on the influences linked with varying operational parameters and fouling of the membrane. As an outcome, one way for decreasing the concentration of the extracellular polymer substances, and therefore decreasing the fouling of membrane, is to increase HRT and SRT. It must be noted, though, that mechanism of membrane fouling in AnMBRs might be comparable to those of the aerobic MBRs but the nature of foulants is different.

Precipitation of struvite and inorganic salts in AnMBR, for instance, inclines to toughen the fouling cake, quickly increasing the fouling rate. More studies are also needed on the adding of coagulants as a means of decreasing membrane fouling, in aerobic and anaerobic systems (Smith et al., 2014).

There also exists the scope for improving the reactor or design of the membrane chamber and enhancing the gas scouring procedure as a means of permanently decreasing membrane fouling.

8.7.2 Membrane Surface Modification

In spite of substantial progress in the growth of the membrane materials over current decades, membranes are still responsible for a substantial part of reactor capital expenses, mainly because of fouling and the inadequate lifetime of existent membranes. Hence, struggles are still required on decreasing membrane fouling, for instance through adjusting commercial membrane surfaces or making new membrane materials, like the low-cost filters or dynamic and the self-assembly membranes (Martin et al., 2011).

Minimization of the fouling of the membrane can be accomplished through alteration of the surface of the membrane. Various modification techniques, like the utilization of nanotechnology or preparation of the new membrane kinds with extra surface properties, have been made and tested till now. Generally speaking, these all methods are all centered on altering physical or chemical interaction with the surface of the membrane, subsequent in an upsurge in the hydrophilic surface character, for instance: decreasing hydrophobic interactions amongst the surface of membrane and compounds or microorganisms present in suspension. As a direct outcome, foulants dropped on the membrane are more readily eliminated during cleaning.

Chemical modification approaches, for instance, covalent bonding or self-assembly approach yield a thin film on the surface of the membrane. A variety of compounds have been utilized for these techniques, comprising silver and the silver nanoparticles, polyvinyl alcohol, polyethylene oxide, zirconium compounds, and titanium or magnesium oxide, all confer extra properties to the surface of the membrane. Physical modification techniques comprise, for instance, plasmatic membrane surface modification utilizing nitrogen, ammonia or air and carbon dioxide. Gamma irradiation and ultra-violet light have also been utilized physically to modify the surface properties of the membrane. The main disadvantage of such kind of high-energy methods, conversely, is the substantial increase in the final membrane production expenses (Gabarrón et al., 2014).

8.7.3 Application of Cheap and Substitute Membrane Materials

Use of cheap and substitute membrane materials not only outcomes in the capital savings but can decrease the operational expenses of the membrane unit. Several studies have scrutinized the utilization of substitute membrane materials, the outcomes of which specify no substantial loss in the effluent quality than that produced using ceramic or polymer membranes. Ferrer et al. (2015) for instance, stated COD elimination efficiencies of up to 91percent in an aerobic/anaerobic bioreactor coupled with the non-woven module. Similarly, Cho et al. (2005) described an insignificant difference in the quality of effluent in the non-woven membrane system than that utilizing the polymer hollow-fiber membrane. Most materials used, till now, conversely, have been verified under laboratory-scale conditions or in the aerobic MBRs. There is a necessity, thus, for studies evaluating the performance of such kind of materials in full or pilot scale AnMBRs (Ramos et al., 2014).

8.7.4 Dynamic and Self-assembly Membranes

Dynamic membranes seem to be an auspicious approach for solving issues linked with high membrane price, quick membrane fouling and the low membrane flux as such kind of membranes display low price with high permeation at the low trans-membrane pressure. Furthermore, dynamic membranes can improve the quality of effluent (Kim et al., 2007).

Dynamic membranes can yield effluent of very high quality; conversely, effluent characteristics rely much on the surrounding conditions near to the membrane. Therefore, dynamic membranes must be kept under optimum conditions through their operation. As the size of a pore of such kind of membranes is highly adjustable, preservation of some compounds and particles can be poorer compared to when using ceramic or polymer membranes. Furthermore, effluent quality can alter over time because of changes in the operational conditions. In contrast, if permeate quality isn't the key criterion, its inferior quality can be balanced by a higher hydraulic performance of the dynamic membranes, together with the lowered operational costs. Additionally, effluents from the anaerobic systems normally suffer a further post-treatment step to enhance their quality before discharge into the receiving water body (Choo & Lee, 1996; Choo et al., 2000).

REFERENCES

1. Al-Hashimia, M. A. I., Abbas, T. R., & Jasema, Y. I. (2013). Performance of sequencing anoxic/anaerobic membrane bioreactor (SAM) system in hospital wastewater treatment and reuse. *European Scientific Journal*, *9*(15), 1-33.
2. Álvarez, J. A., Ruiz, I., Gómez, M., Presas, J., & Soto, M. (2006). Start-up alternatives and performance of an UASB pilot plant treating diluted municipal wastewater at low temperature. *Bioresource Technology*, *97*(14), 1640-1649.
3. Andrade, L. H. D., Mendes, F. D. D. S., Espindola, J. C., & Amaral, M. C. S. (2014). Internal versus external submerged membrane bioreactor configurations for dairy wastewater treatment. *Desalination and Water Treatment*, *52*(16-18), 2920-2932.
4. Aquino, S. F., Hu, A. Y., Akram, A., & Stuckey, D. C. (2006). Characterization of dissolved compounds in submerged anaerobic membrane bioreactors (SAMBRs). *Journal of Chemical Technology & Biotechnology: International Research in Process, Environmental & Clean Technology*, *81*(12), 1894-1904.
5. Araya, P., Aroca, G., & Chamy, R. (1999). Anaerobic treatment of effluents from an industrial polymers synthesis plant. *Waste Management*, *19*(2), 141-146.
6. Aslan, M., Saatçi, Y., Hanay, Ö., & Hasar, H. (2014). Membrane fouling control in anaerobic submerged membrane bioreactor. *Desalination and Water Treatment*, *52*(40-42), 7520-7530.
7. Blaszczyk, R., Gardner, D., & Kosaric, N. (1994). Response and recovery of anaerobic granules from shock loading. *Water Research*, *28*(3), 675-680.
8. Bornare, J. B., Raman, V. K., Sapkal, V. S., Sapkal, R. S., Minde, G., & Sapkal, P. V. (2014). An overview of membrane bioreactors for anaerobic treatment of wastewaters. *Int. J. In. Res. Ad. Eng*, *1*, 91-97.
9. Boyle-Gotla, A., Jensen, P. D., Yap, S. D., Pidou, M., Wang, Y., & Batstone, D. J. (2014). Dynamic multidimensional modelling of submerged membrane bioreactor fouling. *Journal of membrane science*, *467*, 153-161.
10. Chan, Y. J., Chong, M. F., Law, C. L., & Hassell, D. G. (2009). A review on anaerobic–aerobic treatment of industrial and municipal wastewater. *Chemical Engineering Journal*, *155*(1-2), 1-18.

11. Cho, J., Song, K. G., Lee, S. H., & Ahn, K. H. (2005). Sequencing anoxic/anaerobic membrane bioreactor (SAM) pilot plant for advanced wastewater treatment. *Desalination*, *178*(1-3), 219-225.
12. Choi, W. H., Shin, C. H., Son, S. M., Ghorpade, P. A., Kim, J. J., & Park, J. Y. (2013). Anaerobic treatment of palm oil mill effluent using combined high-rate anaerobic reactors. *Bioresource technology*, *141*, 138-144.
13. Choo, K. H., & Lee, C. H. (1996). Membrane fouling mechanisms in the membrane-coupled anaerobic bioreactor. *Water research*, *30*(8), 1771-1780.
14. Choo, K. H., Kang, I. J., Yoon, S. H., Park, H., Kim, J. H., Adiya, S., & Lee, C. H. (2000). Approaches to membrane fouling control in anaerobic membrane bioreactors. *Water science and technology*, *41*(10-11), 363-371.
15. Christian, S., Grant, S., McCarthy, P., Wilson, D., & Mills, D. (2011). The first two years of full-scale anaerobic membrane bioreactor (AnMBR) operation treating high-strength industrial wastewater. *Water Practice and Technology*, *6*(2), 1-44.
16. Cote, P., & Thompson, D. (2000). Wastewater treatment using membranes: the North American experience. *Water science and technology*, *41*(10-11), 209-215.
17. Couras, C. S., Louros, V. L., Grilo, A. M., Leitão, J. H., Capela, M. I., Arroja, L. M., & Nadais, M. H. (2014). Effects of operational shocks on key microbial populations for biogas production in UASB (Upflow Anaerobic Sludge Blanket) reactors. *Energy*, *73*, 866-874.
18. De Vrieze, J., Hennebel, T., Van den Brande, J., Ro'il, M. B., Bruton, T. A., Vankelecom, I. F., ... & Boon, N. (2014). Anaerobic digestion of molasses by means of a vibrating and non-vibrating submerged anaerobic membrane bioreactor. *biomass and bioenergy*, *68*, 95-105.
19. Deowan, S. A., Bouhadjar, S. I., & Hoinkis, J. (2015). Membrane bioreactors for water treatment. In *Advances in Membrane Technologies for Water Treatment* (Vol.1, pp. 155-184). Woodhead Publishing.
20. Dereli, R. K., Ersahin, M. E., Ozgun, H., Ozturk, I., Jeison, D., van der Zee, F., & van Lier, J. B. (2012). Potentials of anaerobic membrane bioreactors to overcome treatment limitations induced by industrial wastewaters. *Bioresource Technology*, *122*, 160-170.
21. Dereli, R. K., Heffernan, B., Grelot, A., van der Zee, F. P., & van Lier,

J. B. (2015). Influence of high lipid containing wastewater on filtration performance and fouling in AnMBRs operated at different solids retention times. *Separation and Purification Technology*, *139*, 43-52.

22. Dereli, R. K., van der Zee, F. P., Heffernan, B., Grelot, A., & van Lier, J. B. (2014). Effect of sludge retention time on the biological performance of anaerobic membrane bioreactors treating corn-to-ethanol thin stillage with high lipid content. *Water research*, *49*, 453-464.
23. Diez, V., Ramos, C., & Cabezas, J. L. (2012). Treating wastewater with high oil and grease content using an Anaerobic Membrane Bioreactor (AnMBR). Filtration and cleaning assays. *Water Science and Technology*, *65*(10), 1847-1853.
24. Dos Santos, A. B., Cervantes, F. J., & van Lier, J. B. (2007). Review paper on current technologies for decolourisation of textile wastewaters: perspectives for anaerobic biotechnology. *Bioresource technology*, *98*(12), 2369-2385.
25. Dvořák, L., Gomez, M., Dvořáková, M., Růžičková, I., & Wanner, J. (2011). The impact of different operating conditions on membrane fouling and EPS production. *Bioresource technology*, *102*(13), 6870-6875.
26. Dvořák, L., Lederer, T., Jirků, V., Masak, J., & Novak, L. (2014). Removal of aniline, cyanides and diphenylguanidine from industrial wastewater using a full-scale moving bed biofilm reactor. *Process Biochemistry*, *49*(1), 102-109.
27. Ellouze, M., Saddoud, A., Dhouib, A., & Sayadi, S. (2009). Assessment of the impact of excessive chemical additions to municipal wastewaters and comparison of three technologies in the removal performance of pathogens and toxicity. *Microbiological research*, *164*(2), 138-148.
28. Ersu, C. B., & Ong, S. K. (2008). Treatment of wastewater containing phenol using a tubular ceramic membrane bioreactor. *Environmental technology*, *29*(2), 225-234.
29. Fenu, A., Roels, J., Wambecq, T., De Gussem, K., Thoeye, C., De Gueldre, G., & Van De Steene, B. (2010). Energy audit of a full scale MBR system. *Desalination*, *262*(1-3), 121-128.
30. Ferrer, J., Pretel, R., Durán, F., Giménez, J. B., Robles, A., Ruano, M. V., ... & Seco, A. (2015). Design methodology for submerged anaerobic membrane bioreactors (AnMBR): A case study. *Separation*

and Purification Technology, *141*, 378-386.

31. Gabarrón, S., Ferrero, G., Dalmau, M., Comas, J., & Rodriguez-Roda, I. (2014). Assessment of energy-saving strategies and operational costs in full-scale membrane bioreactors. *Journal of environmental management*, *134*, 8-14.
32. Gabarrón, S., Gómez, M., Dvořák, L., Růžičková, I., Rodriguez-Roda, I., & Comas, J. (2014). Ragging in mbr: Effects of operational conditions, chemical cleaning, and pre-treatment improvements. *Separation Science and Technology*, *49*(14), 2115-2123.
33. Gao, W. J., Lin, H. J., Leung, K. T., & Liao, B. Q. (2010). Influence of elevated pH shocks on the performance of a submerged anaerobic membrane bioreactor. *Process Biochemistry*, *45*(8), 1279-1287.
34. Ho, J., & Sung, S. (2010). Methanogenic activities in anaerobic membrane bioreactors (AnMBR) treating synthetic municipal wastewater. *Bioresource technology*, *101*(7), 2191-2196.
35. Hu, A. Y., & Stuckey, D. C. (2006). Treatment of dilute wastewaters using a novel submerged anaerobic membrane bioreactor. *Journal of environmental engineering*, *132*(2), 190-198.
36. Ivanovic, I., & Leiknes, T. O. (2012). The biofilm membrane bioreactor (BF-MBR)—a review. *Desalination and Water Treatment*, *37*(1-3), 288-295.
37. Jensen, P. D., Yap, S. D., Boyle-Gotla, A., Janoschka, J., Carney, C., Pidou, M., & Batstone, D. J. (2015). Anaerobic membrane bioreactors enable high rate treatment of slaughterhouse wastewater. *Biochemical Engineering Journal*, *97*, 132-141.
38. Jeong, E., Kim, H. W., Nam, J. Y., Ahn, Y. T., & Shin, H. S. (2010). Effects of the hydraulic retention time on the fouling characteristics of an anaerobic membrane bioreactor for treating acidifi ed wastewater. *Desalination and Water Treatment*, *18*(1-3), 251-256.
39. Judd, S. (2010). *The MBR book: principles and applications of membrane bioreactors for water and wastewater treatment (Vol.1, pp. 1-44)*. Elsevier.
40. Kanai, M., Ferre, V., Wakahara, S., Yamamoto, T., & Moro, M. (2010). A novel combination of methane fermentation and MBR—Kubota Submerged Anaerobic Membrane Bioreactor process. *Desalination*, *250*(3), 964-967.
41. Kijjanapanich, P., Do, A. T., Annachhatre, A. P., Esposito, G., Yeh, D. H.,

& Lens, P. N. (2014). Biological sulfate removal from construction and demolition debris leachate: effect of bioreactor configuration. *Journal of hazardous materials*, *269*, 38-44.

42. Kim, J., Lee, C. H., & Choo, K. H. (2007). Control of struvite precipitation by selective removal of $${\text {NH}}^{{\text {+}}} _ {{\text {4}}} $$ with dialyzer/zeolite in an anaerobic membrane bioreactor. *Applied microbiology and biotechnology*, *75*(1), 187-193.
43. Le-Clech, P. (2010). Membrane bioreactors and their uses in wastewater treatments. *Applied microbiology and biotechnology*, *88*(6), 1253-1260.
44. Lesjean, B., Rosenberger, S., Laabs, C., Jekel, M., Gnirss, R., & Amy, G. (2005). Correlation between membrane fouling and soluble/colloidal organic substances in membrane bioreactors for municipal wastewater treatment. *Water science and technology*, *51*(6-7), 1-8.
45. Lew, B., Tarre, S., Beliavski, M., Dosoretz, C., & Green, M. (2009). Anaerobic membrane bioreactor (AnMBR) for domestic wastewater treatment. *Desalination*, *243*(1-3), 251-257.
46. Liao, B. Q., Kraemer, J. T., & Bagley, D. M. (2006). Anaerobic membrane bioreactors: applications and research directions. *Critical Reviews in Environmental Science and Technology*, *36*(6), 489-530.
47. Lin, H. J., Xie, K., Mahendran, B., Bagley, D. M., Leung, K. T., Liss, S. N., & Liao, B. Q. (2009). Sludge properties and their effects on membrane fouling in submerged anaerobic membrane bioreactors (SAnMBRs). *Water Research*, *43*(15), 3827-3837.
48. Lin, H., Liao, B. Q., Chen, J., Gao, W., Wang, L., Wang, F., & Lu, X. (2011). New insights into membrane fouling in a submerged anaerobic membrane bioreactor based on characterization of cake sludge and bulk sludge. *Bioresource technology*, *102*(3), 2373-2379.
49. Lin, H., Peng, W., Zhang, M., Chen, J., Hong, H., & Zhang, Y. (2013). A review on anaerobic membrane bioreactors: applications, membrane fouling and future perspectives. *Desalination*, *314*, 169-188.
50. Martin, I., Pidou, M., Soares, A., Judd, S., & Jefferson, B. (2011). Modelling the energy demands of aerobic and anaerobic membrane bioreactors for wastewater treatment. *Environmental technology*, *32*(9), 921-932.
51. Martinez-Sosa, D., Helmreich, B., Netter, T., Paris, S., Bischof, F., & Horn, H. (2011). Anaerobic submerged membrane bioreactor

(AnSMBR) for municipal wastewater treatment under mesophilic and psychrophilic temperature conditions. *Bioresource technology*, *102*(22), 10377-10385.

52. Meabe, E., Déléris, S., Soroa, S., & Sancho, L. (2013). Performance of anaerobic membrane bioreactor for sewage sludge treatment: Mesophilic and thermophilic processes. *Journal of membrane science*, *446*, 26-33.
53. Meng, F., Chae, S. R., Drews, A., Kraume, M., Shin, H. S., & Yang, F. (2009). Recent advances in membrane bioreactors (MBRs): membrane fouling and membrane material. *Water research*, *43*(6), 1489-1512.
54. Meng, F., Shi, B., Yang, F., & Zhang, H. (2007). Effect of hydraulic retention time on membrane fouling and biomass characteristics in submerged membrane bioreactors. *Bioprocess and Biosystems Engineering*, *30*(5), 359-367.
55. Minami, K. (1994). A trial of high performance anaerobic treatment on wastewater from a kraft pulp mill. *Desalination*, *98*(1-3), 273-283.
56. Mota, V. T., Santos, F. S., & Amaral, M. C. (2013). Two-stage anaerobic membrane bioreactor for the treatment of sugarcane vinasse: assessment on biological activity and filtration performance. *Bioresource technology*, *146*, 494-503.
57. Ng, K. K., Shi, X., Tang, M. K. Y., & Ng, H. Y. (2014). A novel application of anaerobic bio-entrapped membrane reactor for the treatment of chemical synthesis-based pharmaceutical wastewater. *Separation and Purification Technology*, *132*, 634-643.
58. Ozgun, H., Dereli, R. K., Ersahin, M. E., Kinaci, C., Spanjers, H., & van Lier, J. B. (2013). A review of anaerobic membrane bioreactors for municipal wastewater treatment: integration options, limitations and expectations. *Separation and Purification Technology*, *118*, 89-104.
59. Pandey, A., & Singh, R. K. (2014). Industrial waste water treatment by membrane bioreactor system. *Elixir Chem. Eng*, *70*, 23772-23777.
60. Pretel, R., Robles, A., Ruano, M. V., Seco, A., & Ferrer, J. (2014). The operating cost of an anaerobic membrane bioreactor (AnMBR) treating sulphate-rich urban wastewater. *Separation and Purification Technology*, *126*, 30-38.
61. Qiao, W., Takayanagi, K., Shofie, M., Niu, Q., Yu, H. Q., & Li, Y. Y. (2013). Thermophilic anaerobic digestion of coffee grounds with and without waste activated sludge as co-substrate using a submerged

AnMBR: System amendments and membrane performance. *Bioresource technology*, *150*, 249-258.

62. Qiao, W., Takayanagi, K., Shofie, M., Niu, Q., Yu, H. Q., & Li, Y. Y. (2013). Thermophilic anaerobic digestion of coffee grounds with and without waste activated sludge as co-substrate using a submerged AnMBR: System amendments and membrane performance. *Bioresource technology*, *150*, 249-258.
63. Ramos, C., García, A., & Diez, V. (2014). Performance of an AnMBR pilot plant treating high-strength lipid wastewater: biological and filtration processes. *Water research*, *67*, 203-215.
64. Ramos, C., Zecchino, F., Ezquerra, D., & Diez, V. (2014). Chemical cleaning of membranes from an anaerobic membrane bioreactor treating food industry wastewater. *Journal of membrane science*, *458*, 179-188.
65. Saddoud, A., Ellouze, M., Dhouib, A., & Sayadi, S. (2006). A comparative study on the anaerobic membrane bioreactor performance during the treatment of domestic wastewaters of various origins. *Environmental technology*, *27*(9), 991-999.
66. Santos, A., & Judd, S. (2010). The commercial status of membrane bioreactors for municipal wastewater. *Separation Science and Technology*, *45*(7), 850-857.
67. Savant, D. V., Abdul-Rahman, R., & Ranade, D. R. (2006). Anaerobic degradation of adsorbable organic halides (AOX) from pulp and paper industry wastewater. *Bioresource Technology*, *97*(9), 1092-1104.
68. Seghezzo, L., Zeeman, G., van Lier, J. B., Hamelers, H. V. M., & Lettinga, G. (1998). A review: the anaerobic treatment of sewage in UASB and EGSB reactors. *Bioresource technology*, *65*(3), 175-190.
69. Skouteris, G., Hermosilla, D., López, P., Negro, C., & Blanco, Á. (2012). Anaerobic membrane bioreactors for wastewater treatment: a review. *Chemical Engineering Journal*, *198*, 138-148.
70. Smith, A. L., Stadler, L. B., Cao, L., Love, N. G., Raskin, L., & Skerlos, S. J. (2014). Navigating wastewater energy recovery strategies: a life cycle comparison of anaerobic membrane bioreactor and conventional treatment systems with anaerobic digestion. *Environmental science & technology*, *48*(10), 5972-5981.
71. Spagni, A., Casu, S., Crispino, N. A., Farina, R., & Mattioli, D. (2010). Filterability in a submerged anaerobic membrane

bioreactor. *Desalination*, *250*(2), 787-792.

72. Speece, R. E. (1983). Anaerobic biotechnology for industrial wastewater treatment. *Environmental science & technology*, *17*(9), 416A-427A.
73. Stuckey, D. C. (2012). Recent developments in anaerobic membrane reactors. *Bioresource technology*, *122*, 137-148.
74. Tao, Y., Gao, D. W., Fu, Y., Wu, W. M., & Ren, N. Q. (2012). Impact of reactor configuration on anammox process start-up: MBR versus SBR. *Bioresource Technology*, *104*, 73-80.
75. Van Lier, J. B. (2008). High-rate anaerobic wastewater treatment: diversifying from end-of-the-pipe treatment to resource-oriented conversion techniques. *Water Science and Technology*, *57*(8), 1137-1148.
76. Van Lier, J. B. (2008). High-rate anaerobic wastewater treatment: diversifying from end-of-the-pipe treatment to resource-oriented conversion techniques. *Water Science and Technology*, *57*(8), 1137-1148.
77. Vyrides, I., & Stuckey, D. C. (2009). Saline sewage treatment using a submerged anaerobic membrane reactor (SAMBR): effects of activated carbon addition and biogas-sparging time. *Water research*, *43*(4), 933-942.
78. Wallace, J. M., & Safferman, S. I. (2014). Anaerobic membrane bioreactors and the influence of space velocity and biomass concentration on methane production for liquid dairy manure. *Biomass and Bioenergy*, *66*, 143-150.
79. Wang, W., Yang, Q., Zheng, S., & Wu, D. (2013). Anaerobic membrane bioreactor (AnMBR) for bamboo industry wastewater treatment. *Bioresource technology*, *149*, 292-300.
80. Wei, C. H., Harb, M., Amy, G., Hong, P. Y., & Leiknes, T. (2014). Sustainable organic loading rate and energy recovery potential of mesophilic anaerobic membrane bioreactor for municipal wastewater treatment. *Bioresource technology*, *166*, 326-334.
81. Wijekoon, K. C., Visvanathan, C., & Abeynayaka, A. (2011). Effect of organic loading rate on VFA production, organic matter removal and microbial activity of a two-stage thermophilic anaerobic membrane bioreactor. *Bioresource Technology*, *102*(9), 5353-5360.
82. Wong, K., Xagoraraki, I., Wallace, J., Bickert, W., Srinivasan, S., & Rose, J. B. (2009). Removal of viruses and indicators by anaerobic

membrane bioreactor treating animal waste. *Journal of environmental quality*, *38*(4), 1694-1699.

83. Xie, Z., Wang, Z., Wang, Q., Zhu, C., & Wu, Z. (2014). An anaerobic dynamic membrane bioreactor (AnDMBR) for landfill leachate treatment: performance and microbial community identification. *Bioresource technology*, *161*, 29-39.
84. Yu, L., Zhang, Y., Zhang, B., Liu, J., Zhang, H., & Song, C. (2013). Preparation and characterization of HPEI-GO/PES ultrafiltration membrane with antifouling and antibacterial properties. *Journal of membrane science*, *447*, 452-462.
85. Zhang, X., Yue, X., Liu, Z., Li, Q., & Hua, X. (2015). Impacts of sludge retention time on sludge characteristics and membrane fouling in a submerged anaerobic–oxic membrane bioreactor. *Applied microbiology and biotechnology*, *99*(11), 4893-4903.
86. Zsirai, T., Buzatu, P., Aerts, P., & Judd, S. (2012). Efficacy of relaxation, backflushing, chemical cleaning and clogging removal for an immersed hollow fibre membrane bioreactor. *Water research*, *46*(14), 4499-4507.

INDEX

A

S

T

U

V

W